Springer Theses

Recognizing Outstanding Ph.D. Research

Aims and Scope

The series "Springer Theses" brings together a selection of the very best Ph.D. theses from around the world and across the physical sciences. Nominated and endorsed by two recognized specialists, each published volume has been selected for its scientific excellence and the high impact of its contents for the pertinent field of research. For greater accessibility to non-specialists, the published versions include an extended introduction, as well as a foreword by the student's supervisor explaining the special relevance of the work for the field. As a whole, the series will provide a valuable resource both for newcomers to the research fields described, and for other scientists seeking detailed background information on special questions. Finally, it provides an accredited documentation of the valuable contributions made by today's younger generation of scientists.

Theses may be nominated for publication in this series by heads of department at internationally leading universities or institutes and should fulfill all of the following criteria

- They must be written in good English.
- The topic should fall within the confines of Chemistry, Physics, Earth Sciences, Engineering and related interdisciplinary fields such as Materials, Nanoscience, Chemical Engineering, Complex Systems and Biophysics.
- The work reported in the thesis must represent a significant scientific advance.
- If the thesis includes previously published material, permission to reproduce this must be gained from the respective copyright holder (a maximum 30% of the thesis should be a verbatim reproduction from the author's previous publications).
- They must have been examined and passed during the 12 months prior to nomination.
- Each thesis should include a foreword by the supervisor outlining the significance of its content.
- The theses should have a clearly defined structure including an introduction accessible to new PhD students and scientists not expert in the relevant field.

Indexed by zbMATH.

Thomas Groetsch

Enhancing Carbon Fibre Production Through Life Cycle Analysis

Doctoral Thesis accepted by
Deakin University, Waurn Ponds, Australia

Springer

Author
Dr. Thomas Groetsch
Institute for Frontier Materials
Deakin University
Waurn Ponds, VIC, Australia

Supervisor
Prof. Russell Varley
Institute for Frontier Materials
Deakin University
Waurn Ponds, VIC, Australia

ISSN 2190-5053 ISSN 2190-5061 (electronic)
Springer Theses
ISBN 978-3-031-70577-9 ISBN 978-3-031-70578-6 (eBook)
https://doi.org/10.1007/978-3-031-70578-6

This Springer imprint is published by the registered company Springer Nature Switzerland AG
The registered company address is: Gewerbestrasse 11, 6330 Cham, Switzerland

Supervisor's Foreword

Life cycle assessment (LCA) is critical to fully evaluate the environmental impact of products and processes. As the world moves towards a more responsible use of resources through recycling and more bio-based alternatives to fossil fuels, effective implementation of assessment tools is vital. Carbon fibre, the material of the twenty-first century, due to its lightweight and high strength, provides solutions to reducing greenhouse gas emissions through increased fuel efficiencies and contributes to lower-cost renewable energy. However, understanding the overall environmental benefit, by considering the energy consumption during its manufacture is just as important as the environmental impact of the end-use applications. Therefore, to understand the true environmental cost of carbon fibre, a streamlined analytical model tailored to carbon fibre manufacture was developed. The gate-to-gate assessment concept was employed to perform a detailed analysis of the manufacturing process using different precursor materials on a large production scale. Due to the proprietary nature of the carbon fibre industry, most publicly available information about the material and production process is based on laboratory-scale research, assessing the economic and environmental impact of carbon fibre production impossible. This research addresses the significant research gap in available data for large, continuous-scale manufacturing, representative of industrial carbon fibre processes.

The modular LCA concept used throughout this work adheres to international life cycle assessment standards and incorporates a limited life cycle costing perspective to account for the high cost and value of the material being investigated. The methodology was designed to efficiently generate standardized results focussing on key process steps and parameters, thus enabling streamlined comparison of varying process or material scenarios.

The economic and environmental assessment concept was applied to analyse the carbon fibre manufacturing process for a commercial polyacrylonitrile (PAN) precursor on an industrially relevant manufacturing plant. The process analysis focussed on material in- and output in each production step and revealed for the first time a detailed breakdown of emissions generated in each section of the process.

This in-depth process analysis was further employed to compare processing scenarios designed for higher material throughput. The analysis identified an operating window in which the production process can be optimized, limited by detrimental effects on materials properties.

In order to assess the economic and environmental impact of precursor on the carbon fibre manufacturing process, three different precursor materials were examined. The gate-to-gate analysis revealed precursor-specific advantages and disadvantages. While the evaluation of traditional carbon fibre precursors highlighted the importance of processability for achieving advanced material properties, the assessment of bio-derived precursor material emphasized the cost advantage of the abundantly available source material.

Overall, this work highlights key areas of interest from an environmental and economical perspective when developing improved processes or precursor materials, such as higher throughput and more sustainable material solutions. Additionally, by correlating process parameters, material properties, environmental implications and economic factors, this work contributes to a more holistic picture of the carbon fibre manufacturing process.

All research presented in this book was conducted at Carbon Nexus at Deakin University, Australia, the only open-access carbon fibre manufacturing facility at the necessary scale. The streamlined nature of the presented LCA methodology significantly enhances the quality and detail of information available with respect to the economic and environmental impact associated with source materials and continuous carbon fibre manufacture.

Waurn Ponds, Australia
May 2023

Prof. Russell Varley

Parts of this thesis have been published in the following articles:
Refereed Journal Papers

1. **T. Groetsch**, C. Creighton, R. Varley, A. Kaluza, A. Der, F. Cerdas, C. Hermann
   "*A modular LCA/LCC-modelling concept for evaluating material and process innovations in carbon fibre manufacturing*"
   Procedia CIRP, Volume 98, 2021, Pages 529–534,
   https://doi.org/10.1016/j.procir.2021.01.146
2. **T. Groetsch**, M. Maghe, R. Rana, R. Hess, S. Nunna, J. Herron, D. Buckmaster, C. Creighton, R. Varley
   "*Gas Emission Study of the Polyacrylonitrile-Based Continuous Pilot-Scale Carbon Fiber Manufacturing Process*"
   Industrial & Engineering Chemistry Research, Volume 60, Issue 48, 2021, Pages 17379–17389
   https://doi.org/10.1021/acs.iecr.1c02253
3. **T. Groetsch**, M. Maghe, C. Creighton, R. Varley
   "*Environmental, property and cost impact analysis of carbon fibre at increasing rates of production*"
   Journal of Cleaner Production, Volume 382, 2023, 135292
   https://doi.org/10.1016/j.jclepro.2022.135292
4. **T. Groetsch**, M. Maghe, C. Creighton, R. Varley
   "*Economic and environmental effects of precursor variation in a continuous carbon fibre manufacturing process*"
   Journal of Industrial and Engineering Chemistry, Volume 127, 2023, Pages 554–566
   https://doi.org/10.1016/j.jiec.2023.07.041

Acknowledgements

Firstly, I want to thank my principal supervisor, Dr. Claudia Creighton, for giving me the opportunity to work on this project, providing valuable guidance, always constructive and energizing feedback, and for the countless discussions over the last few years that greatly improved this work. I also would like to thank my supervisor, Prof. Russell Varley, for supporting me with valuable feedback, patience, knowledge, and a well-grounded perspective. I also want to thank my external supervisor, Prof. Christoph Herrmann, and his team at TU Braunschweig for the continuous support, discussions, and feedback on all matters related to LCA.

I would like to thank Dr. Maxime Maghe for sharing his knowledge and ideas, and for helping me design, conduct, and analyse much of the experimental part of this work. I also would like to express my gratitude to John Herron and Scott Anderson for never denying my countless requests to collect more samples or run additional trials, for sharing their knowledge, and always providing solutions for any production-related challenges that occurred. I also would like to thank the wider Carbon Nexus team for their support and for always having an open ear for any questions or requests.

I would like to acknowledge Deakin University for the financial support during this project through a DUPR scholarship.

To my sister Julia, whose love for non-cutting manufacturing techniques has always been an inspiration.

I would like to express my sincere gratitude to my parents. I owe them a debt greater than I will ever be able to repay. All that I am, I owe to them.

And finally, a special thanks to my wife Melissa for filling my life with happiness, for calmly sitting through countless tirades when experiments just did not go the way they were supposed to and for offering encouragement when it was most needed.

Contents

Chapter 1
Introduction

Global manufacturing industries today face many challenges due to changing legislation, increasing standards and intensifying international competition. In many instances, these changes reflect the need to address challenges related to global warming but also rising raw material prices [1].

1.1 Research Field and Context

One of the main drivers for more restrictive laws and standards is the intention to reduce the effect of the changing climate worldwide [2, 3]. Research and the resulting legislation often focus on CO_2 emission as a general measure for comparing the effect of technology, industrial processes or energy production, on the environment. Furthermore recent global developments have highlighted to strong impact of energy supply in all areas of a products life cycle [4]. While these actions are noticeable in many areas, manufacturing industries are particularly affected [5]. As one of the biggest consumers of energy and therefore producer of emissions, legislation is forcing the manufacturing sector to continuously improve its environmental friendliness. For more complex technical products, such as airplanes or automobiles, legislation is driving the development and use of alternative materials that are more sustainable or promise an energy saving benefit over the product lifetime [6]. Carbon fibre reinforced components, for example, offer a significant weight reduction potential, as they are up to 35% lighter than aluminium and 60% lighter than commonly used steel [7]. The reduced weight in turn may decrease energy demand during the use-stage of vehicles for instance. There is, however, a significant hurdle for the widespread success of the use of this material, since the production of carbon fibre consumes substantially more energy than the production of a comparable amount of steel or aluminium, for a product with the closely similar properties [8]. Reducing

T. Groetsch, *Enhancing Carbon Fibre Production Through Life Cycle Analysis*, Springer Theses, https://doi.org/10.1007/978-3-031-70578-6_1

environmental effects and the cost of carbon fibre production is therefore an important factor to increase its usage.

Another problem caused by the constant growth of economies around the world, is the increasing need for raw materials. However, economically exploitable sources of natural resources, such as iron ore and rare-earth elements, are limited. For various materials, the growing demand in combination with declining natural availability leads to bottlenecks and increasing material prices [9, 10]. Plastics and reinforced plastics are increasingly replacing metal parts in various applications, as fossil fuel derived polymer materials are still relatively abundant [11]. However, for applications with high mechanical requirements, metal alloys are still the most commonly used materials, despite the potential for carbon fibre reinforced plastics to replace metals due to their excellent strength to weight ratio. Currently, the main inhibitor for widespread application is the higher price of carbon fibre. However, increasing steel prices and ongoing research towards carbon fibre price reduction, promise a gradual shift towards increased use of carbon fibre reinforced plastic products [9–12].

Carbon fibre composite materials have the potential to play an important role in reducing global greenhouse gas emissions and conservation of fossil resources if sourced from renewable precursor material. In some industries, such as aerospace or wind energy, the adoption of carbon fibres is already widespread, but more price sensitive products, such as automobiles, are still widely manufactured using traditional materials. With growing awareness of climate change and increasingly stringent legislation, the acceptance of carbon fibres in new industry fields is steadily growing. In many areas the use of composite materials, however, is still in an early phase of development or research, but intensifying interest in these products highlight their growing importance [13–16].

The biggest problem of carbon fibre-based material is its high cost, compared to more traditional material like metals or plastics. The price for carbon fibre reinforced components is simply not yet competitive with e.g. steel or aluminium. While improvements have been made over the past years and prices slowly drop, the underlying problem remains. The amount of overall energy and material necessary to produce carbon fibres is still at a very high level. The generation of heat at various stages of the process as well as long cycle times cause high energy demand and therefore high costs. With an overall material yield of around 50% at the fibre production stage, not only is the ratio of material input and output unfavourable, but the lost material is converted into toxic and environmentally detrimental emissions. These gases have to be treated in another energy-intensive step before release into atmosphere. To abate the hazardous gases produced during the process, very high temperatures are necessary, which again drives energy demand and therefore cost. The heat treatment processes can be responsible for over 30% of the total energy demand of carbon fibre production [17]. A lot of research in the field of carbon fibres addresses these problems in one way or another. However, there is still considerable room for improvement, especially in terms of production process parameter optimisation for more sustainable and cost-effective carbon fibre production. As knowledge in regard to processing parameters is usually treated as confidential by the fibre manufacturer, data to analyse and necessary information is only limitedly available [17, 18].

Another issue becoming ever more relevant is the recycling of intermediate or end products containing carbon fibre. As carbon fibres are most commonly used in composites with thermoset resin systems, recovery of the original materials is challenging. Neither the fibres nor the resin system can typically easily be restored to their original conditions, as for example metals can. There are hence various routes suggested by recent research to overcome or bypass some of these issues through reuse, remanufacturing, partial recycling or downcycling [19–21]. With increasing use of carbon fibre composites, recycling strategies and reclaimed carbon fibre will be more widely deployed and available.

1.2 Objective and Research Purpose

This work focused on the analysis of the carbon fibre manufacturing process, evaluation of variations in process parameters and precursor material and identification of improvement potentials using a life cycle analysis-based methodology. Related published research often lacks crucial information, making it difficult to transfer the specific findings to other production settings or different materials. Within this framework, the research aimed at developing and utilizing a streamlined, tailored methodology to evaluate the carbon fibre manufacturing in terms of energy demand, related environmental impact and cost. This furthermore allowed an effective comparison of environmental and economic parameters for the manufacturing process and final product.

Towards this goal, the following work has been completed:

- Development of a streamlined life cycle assessment (LCA) approach tailored to the carbon fibre manufacturing approach focussing on key contributing process steps
- Verification of the concept in an initial assessment with data gathered from the Carbon Nexus manufacturing facilities
- In-depth analysis of the carbon fibre manufacturing process at the Carbon Nexus research line:
  - Process parameter analysis (energy demand, process temperatures, drive speed etc.)
  - In-line emission analysis both on a process step level and overall
  - Material property analysis, single fibre tensile testing, FT-IR analysis, elemental analysis, density, mass density
- Evaluation of influence of process parameter changes on carbon fibre material properties, energy demand, environmental impact and cost
  - Process parameter analysis of four different line speeds
  - Overall in-line emission analysis of four different line speeds
  - Material property analysis, single fibre tensile testing, FT-IR analysis, elemental analysis, density, mass density of four different line speeds.

- Evaluation of influence of precursor material variation on carbon fibre material properties, energy demand, environmental impact and cost
  - Process parameter analysis of two different precursor materials, special grade PAN and textile PAN
  - Overall in-line emission analysis of the two materials
  - Material property analysis, single fibre tensile testing, FT-IR analysis, elemental analysis, density, mass density for the two materials
  - Additional benchtop analysis of a cellulose-lignin blend material as potential alternative and virtual scale-up of results to compare with the other two materials
  - Holistic evaluation of the process parameter changes and material variation findings from a life cycle thinking perspective.

1.3 Outline of Study

The overall structure of this work with four key focus areas is shown in Fig. 1.1.

Chapter 2 gives an overview of the current state of research of life cycle analysis (LCA) and life cycle costing (LCC) methodology development and application focused on carbon fibre research and production, key aspects of the carbon fibre manufacturing process from a life cycle engineering and thinking perspective, research related to process changes and optimisation as well as information about the development and optimization of precursors.

Chapter 3 describes the materials and methodologies applied in this work. It gives an overview of the carbon fibre manufacturing process and key steps from a life cycle engineering and thinking perspective and furthermore outlines the methods used to characterise the processed material and all other analysis techniques employed for this project. Finally, the methodological approach based on LCA and LCC principles is introduced and the streamlining strategies to tailor the approach to carbon fibre processing are outlined.

In Chap. 4 a streamlined LCA/LCC modelling concept was developed. The methodological background of the concept is introduced and relevant content of pertaining standards is outlined. The concept is introduced in detail with detailed descriptions for each step and their linkage to standardized approaches. This section furthermore showcases an exemplary application of the developed methodology, details necessary data to be collected and suggests how to process relevant information.

Chapter 5 presents a study on gas emissions and energy demand during the manufacturing process of carbon fibre based on a PAN precursor, with properties typically used for commercial carbon fibre production. The conversion process is examined with respect to the chemical and mechanical transformation of the material, while energy demand and generated emissions for each major production segment are measured. The aim of this study was to enhance the understanding of relationships

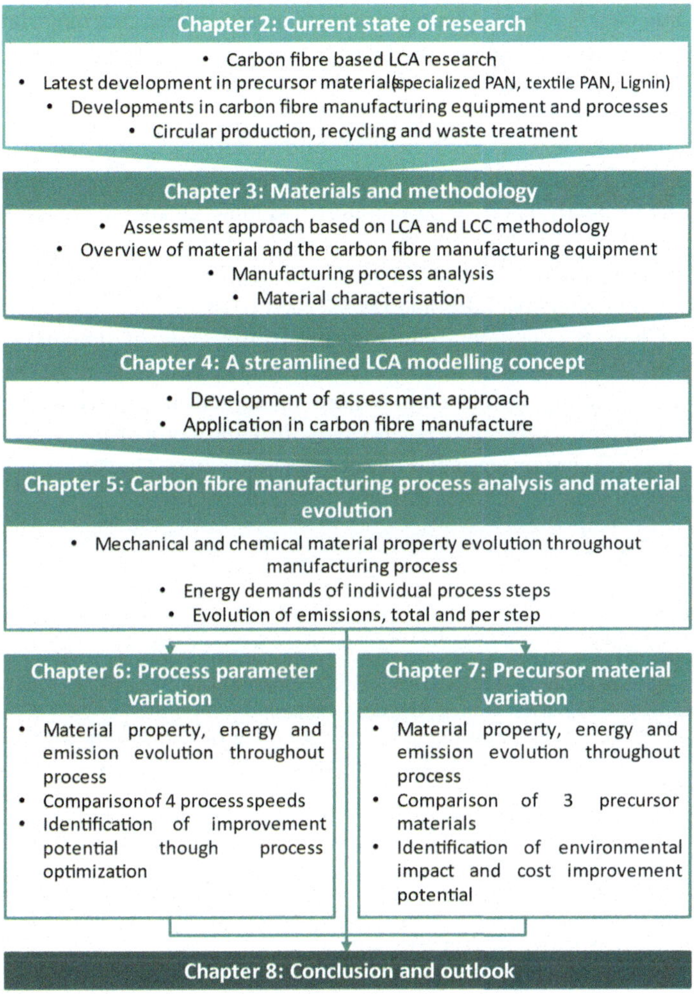

Fig. 1.1 Overview of study structure

between chemical and mechanical changes of the material, energy demand of the various process steps and emissions generated during the transformation.

In Chap. 6 the impact of varying processing speed, namely the residence time in heat, on material transformation, emissions and energy demand was evaluated. Four different process speeds were analysed while the mechanical properties of the resulting carbon fibres were maintained on a closely similar level. This study aimed at identifying pathways to optimize the manufacturing process and to understand which process parameters are the greatest levers to lower energy demand, emissions and cost, without significantly changing the process for a given precursor material.

Expanding on this, Chap. 7 outlines how the choice of precursor material can affect the evolution of chemical and mechanical material properties and related emissions and energy. Two PAN-based precursor materials were processed, one special grade type traditionally used for carbon fibre production and a textile-grade material with a different comonomer composition. The third material investigated is a fully bio-based blend of cellulose and lignin. This material is not processed at the same scale and therefore the analysis relied on literature and material characterisation to enable a comparison. This part of the study aims to establish a more holistic understanding of source-material characteristics that are required to provide a viable alternative to the currently prevalent special-grade PAN.

Finally, Chap. 8 concludes with a general discussion of the findings of this work detailing process- and material-related results and how they contribute to a better understanding of the carbon fibre manufacturing process from a more holistic perspective, encompassing not only structure–property relationships, but also including relevant process parameters and emissions. The key conclusions are summarized and areas of interest for future work are outlined.

References

1. Nachmany M et al (2014) The GLOBE climate legislation study: a review of climate change legislation in 66 countries. United Kingdom, London
2. Ladan MT (2018) Achieving sustainable development goals through effective domestic laws and policies on environment and climate change. Environ Policy Law 48(1):42–63. https://doi.org/10.3233/epl-180049
3. Fawzy S, Osman AI, Doran J, Rooney DW (2020) Strategies for mitigation of climate change: a review. Environ Chem Lett 18(6):2069–2094. https://doi.org/10.1007/s10311-020-01059-w
4. Hu Z (2021) A study of how soaring energy prices affect the economy, 2022-03-16T13:00:00.000Z 2022. In: Proceedings of the 7th international conference on economy, management, law and education (EMLE 2021): Atlantis Press, pp 78–83 [Online]. Available: https://doi.org/10.2991/aebmr.k.220306.012
5. Leong WD, Lam HL, Ng WPQ, Lim CH, Tan CP, Ponnambalam SG (2019) Lean and green manufacturing—a review on its applications and impacts. Process Integration and Optimization for Sustainability 3(1):5–23. https://doi.org/10.1007/s41660-019-00082-x
6. Agarwal J, Sahoo S, Mohanty S, Nayak SK (2019) Progress of novel techniques for lightweight automobile applications through innovative eco-friendly composite materials: a review. J Thermoplast Compos Mater 33(7):978–1013. https://doi.org/10.1177/0892705718815530
7. Das S (2011) Life cycle assessment of carbon fiber-reinforced polymer composites. Int J Life Cycle Assess 16(3):268–282. https://doi.org/10.1007/s11367-011-0264-z
8. Dér A et al (2021) Modelling and analysis of the energy intensity in polyacrylonitrile (PAN) precursor and carbon fibre manufacturing. J Clean Prod 303:127105. https://doi.org/10.1016/j.jclepro.2021.127105
9. George G, Schillebeeckx SJD, Liak TL (2015) The management of natural resources: an overview and research agenda. Acad Manag J 58(6):1595–1613. https://doi.org/10.5465/amj.2015.4006
10. Krautkraemer JA (1998) Nonrenewable resource scarcity. J Econ Lit. 36(4):2065–2107 [Online]. Available: http://www.jstor.org/stable/2565047
11. Kerns J (2016) Replacing metal with plastic. Mach Des 88(9):34–42

12. Font Vivanco D, Sprecher B, Hertwich E (2017) Scarcity-weighted global land and metal footprints. Ecol Indic 83:323–327. https://doi.org/10.1016/j.ecolind.2017.08.004
13. Gupta A, Kumar A (2010) Composites materials: addressing the climate change. Asia Pac Bus Rev 6(1):78–89. https://doi.org/10.1177/097324701000600107
14. Heuss R, Müller N, van Sintern W, Starke A, Tschiesner A (2012) Lightweight, heavy impact—how carbon fiber and other lightweight materials will develop across industries and specifically in automotive. In: Advanced industries
15. Koumoulos EP et al (2019) Research and development in carbon fibers and advanced high-performance composites supply chain in Europe: a roadmap for challenges and the industrial uptake. J Compos Sci 3(3):86. https://doi.org/10.3390/jcs3030086
16. Yu M, Wiedmann T, Crawford R, Tait C (2017) The carbon footprint of Australia's construction sector. Procedia Eng 180:211–220. https://doi.org/10.1016/j.proeng.2017.04.180
17. Arnold U, De Palmenaer A, Brück T, Kuse K (2018) Energy-efficient carbon fiber production with concentrated solar power: process design and techno-economic analysis. Ind Eng Chem Res 57(23):7934–7945. https://doi.org/10.1021/acs.iecr.7b04841
18. Khayyam H et al (2015) Stochastic optimization models for energy management in carbonization process of carbon fiber production. Appl Energy 158:643–655. https://doi.org/10.1016/j.apenergy.2015.08.008
19. Zhang J, Chevali VS, Wang H, Wang C-H (2020) Current status of carbon fibre and carbon fibre composites recycling. Compos B Eng 193:108053. https://doi.org/10.1016/j.compositesb.2020.108053
20. Pimenta S, Pinho ST (2011) Recycling carbon fibre reinforced polymers for structural applications: technology review and market outlook. Waste Manag 31(2):378–392. https://doi.org/10.1016/j.wasman.2010.09.019
21. Pakdel E, Kashi S, Varley R, Wang X (2021) Recent progress in recycling carbon fibre reinforced composites and dry carbon fibre wastes. In: Resources, conservation and recycling, vol 166. https://doi.org/10.1016/j.resconrec.2020.105340

Chapter 2
Literature Review

2.1 Background

With increasing awareness towards the end of the last century, that the development of the world in terms of economy, technology and population needed and still needs to change to support an ever-growing human population, sustainable development and conscious use of natural resources have been increasingly incorporated in economy policies and business strategies. Recent developments on the global energy market, leading to soaring prices and reduced supply of primary energy materials, have further highlighted the need to reduce demand and dependencies on specific resources without compromising the shift towards more sustainable economies [1]. To better understand and address these types of issues across industries, various methods have been proposed over the years, which can be classified under the umbrella term "life cycle engineering" [2]. Specifically focusing on the environmental impact of products in a very broad sense, life cycle assessment (LCA) methods have been developed over the past decade. In the most traditional sense the assessment encompasses every interaction of the product with the environment, direct or indirect, from raw material extraction to the end of product life, disposal or recycling [3]. With globalized industries and highly complex products, this can quickly become an enormous task. With this in mind, the methodology allows for some flexibility adjusting the goal and scope of any study to facilitate concise assessments and reduce complexity.

The almost oligopolistic carbon fibre market entails a very secretive carbon fibre production industry. While fundamental research related to carbon fibre production is widely published and advancements are made continuously in a wide range of fields, little to no information is available on an industrial scale. Due to the high initial capital investment necessary to set up a continuous carbon fibre line, even at a small scale, access to industrially relevant facilities is difficult, with only few open access facilities with such capabilities available worldwide. This again significantly limits publicly available research on anything larger than benchtop scale. For this reason, there are only very few publications available concerning the environmental impact

T. Groetsch, *Enhancing Carbon Fibre Production Through Life Cycle Analysis*, Springer Theses, https://doi.org/10.1007/978-3-031-70578-6_2

of carbon fibre and the materials production. While carbon fibre-based products offer advantages over traditional materials-based goods during use stage, the energy and material intensive production stage and limited recyclability have a detrimental impact on their overall environmental impact. Another aspect related to this is the cost of a product and more general from an LCE-perspective the overall cost of a product throughout its life cycle. Today, no standardized methodology for LCC across industries exists as there is for LCA, however, a general consensus exists as to what an assessment based on this methodology should include. Furthermore standardization of the methodology has begun in certain industries such as he building sector [4]. For a high-value product such as carbon-fibres, it may be beneficial to not only assess the environmental impact but at the same time review the associated costs of the material. Life cycle sustainability assessment (LCSA) encompasses the two approaches while also focusing on social aspects of products and processes [5, 6]. While the research presented here does not include social considerations LCSA can also serve as a reference for the developed methodology.

To address discussed research gap access to both laboratory equipment and pilot scale carbon fibre manufacturing equipment is required. The facilities at Carbon Nexus offer two continuous manufacturing setups at different scales: a smaller research line, capable of processing up to three fibre tows of 24,000 filaments and a bigger pilot line capable of processing up to 28 fibre tows of 24,000 filaments. Both designs fully represent a commercial scale production in terms of process sequence and continuity. The smaller research line has a theoretical yearly output of < 1 ton, while the larger research line could theoretically supply up to approximately 50t per year in a best-case scenario. Commercial production facilities currently used in industry are significantly larger, processing 2000–4000t per year or more [7, 8]. The experimental part of this work relies heavily on the smaller research line, as this enabled extensive variation of parameters and use of different materials. Adjacent laboratories offer direct access to all the necessary material testing equipment outlined in Chap. 3. There may be some scaling effects for larger facilities which may skew results presented here, which has to be considered when assessing shown results and trends.

This chapter outlines the current state of research in the various fields related to the conducted research. The first section gives a very general and short introduction to LCA and a brief overview of current research related to LCA basics. Similarly, LCC is introduced and an overview of recent work in the field is given. As a combination of these two methods was used, research combining methods or more general with a holistic view are discussed, before work more specific to carbon fibre production is covered.

In the second segment common precursor materials, their relevant properties, microstructure and chemistry are introduced. The focus here lies on different types of Polyacrylonitrile (PAN) and bio-based source material such as Lignin and Cellulose. A short overview is also given on previous studies related to alternative materials and blends.

The third part of this review deals with at the carbon fibre production process, with a short overview of the precursor manufacturing process and composite production. This section further includes research pertaining to process optimisation and innovation.

The last section gives an overview of recycling strategies for carbon fibre products and emission treatment for gasses directly generated during the production.

Overall the review discusses the current state of research in the respective areas from an engineering viewpoint with a LCA and cost perspective. In more material related sections, the review aims to focus on material/energy flows, production, environmental impact and cost, with LCA principles in mind.

2.2 Precursor Source Material

The precursor material has great influence on the properties of the produced carbon fibres, production parameters, yield and the overall cost of the final product. There is a wide range of materials available that have the potential to be used as a precursor for carbon fibre production. The most commonly used material in industrial scale production today is Polyacrylonitrile (PAN). With sustainability in mind and the aim to lower cost, research is also focusing on bio-based material, of which lignin and cellulose are currently the most common. There are various other types of precursor materials, both already in use and being developed, e.g. pitch and polyethylene (PE) [9, 10]. Pitch is typically used to produce very high modulus fibres at yield rates as high as 85%, other mechanical properties of these fibres however are typically lower than PAN based material [29]. Most materials are still in very early development phases or have been found to not be commercially viable [9, 10].

2.2.1 Polyacrylonitrile

PAN fibres are currently the predominant precursor type for industrial scale production due to the high yield that can be achieved along with the possible high-performance properties of the final product [9]. PAN is a synthetic polymer with the backbone chemical structure shown in Fig. 2.1. Polyacrylonitrile has a carbon content of 67.9%, which makes it an ideal material to carbonize, as the overall yield achievable is typically above 50% [9, 11]. "Special grade" precursor fibres in use today were developed from material intended for textile products [9]. Despite currently being the most economically viable precursor on industrial scale, the high cost of the material and long necessary processing times significantly contribute to the high market price of carbon fibres [12].

Acrylonitrile monomers are commonly converted to PAN by free radical polymerisation through solution, bulk, emulsion or aqueous dispersion polymerisation

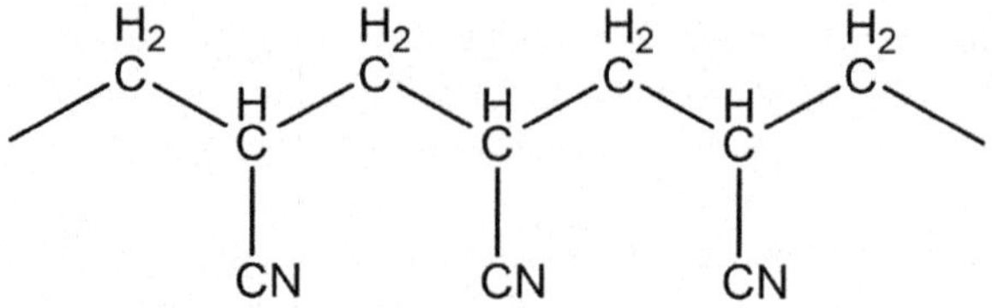

Fig. 2.1 PAN backbone structure

[9, 10, 12]. PAN precursors typically contain at least 85wt% by weight acrylonitrile monomer and up to 15% comonomers correspondingly [9]. Comonomers play a key role in tailoring the properties of the material for the desired application. In a carbon fibre context, PAN homopolymer is challenging to process during the stabilization stage due to a very narrow and strong exothermic reaction window in combination with a high initiation temperature which affects achievable properties [10]. Comonomers can address these issues by lowering the cyclisation initiation temperature and help control the exothermic reaction to achieve better mechanical properties [9, 10]. Comonomers are further used to enhance the solubility, spinnability and improve fibre handling [9]. PAN degrades when it is heated above 180 °C, well before its melting point and therefore cannot easily be melt spun. More complex and expensive wet-spinning techniques, such as solution spinning and air gap spinning, are commonly used to produce fibre material [9, 13]. The high complexity stems from a necessary high process control and advanced equipment to produce high quality, molecularly well oriented fibres with low void-content, while the high process related costs can in part be attributed to the necessary use of toxic organic solvents [12, 13]. The widely used wet spinning process, schematically shown in Fig. 2.2, starts with dissolving the PAN polymer in an organic solvent such as dimethyl sulfoxide (DMSO) or dimethylformamide (DMF) at concentrations between 15 and 20% depending on the polymer and solvent used [9]. This solution is called dope and fed to spinneret at a defined rate, where it enters the coagulation through a number of small holes, a few hundred on experimental setups and above 24,000 for price sensitive industrial precursors [9, 14]. The coagulation bath is typically an aqueous solution, which precipitates the PAN to form continuous fibres at speed of about 3–16 m/min [9]. Subsequently the fibres are washed in hot water and stretched to remove any remaining solvent and align the molecule chains to improve the strength of the material, reduce the size of voids and enable crosslinking of the molecular chains [9, 12]. After a spin finish, acting as lubricant and antistat, is applied, the fibres are dried to remove any remaining liquid. Commonly used spin finishes are aqueous solutions with lubricants such as silicones or esters and antistatic agents such as alkyl compounds or phosphate esters [15]. The elimination of liquids from the inner structures leads to an overall material contraction and a decrease in porosity [9, 12, 14]. Finally the fibres are collected on spools or, for larger tows, crimped in boxes [9].

The wound precursor can then be converted to carbon fibres in a two-stage process with stabilization and carbonization. During the stabilization stage, the initially open

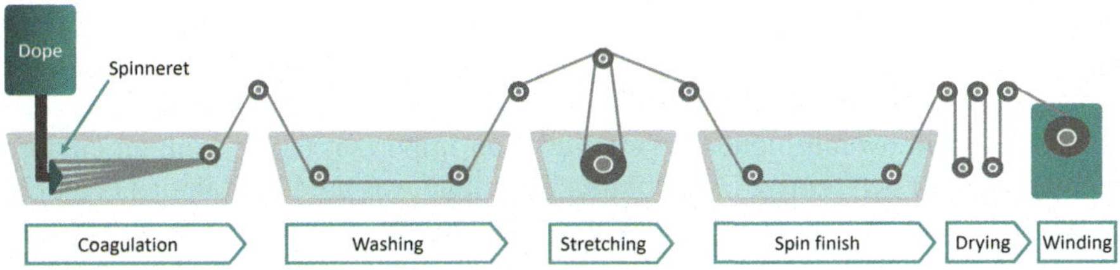

Fig. 2.2 Schematic representation of wet spinning process

chain composition of the material is converted to form ladder-like structures [10, 16]. However, if the heating rate is higher than 5 °C/min the molecular structure is damaged [16]. As the thermal stabilization of the PAN fibres is carried out at 200 °C and above with slowly rising temperatures in air, the process is very time and energy consuming. During the second carbonization stage, temperatures are further increased up to 1500 °C at high heating rates under an inert atmosphere [34]. This part of the manufacturing process will be discussed in greater detail in Sect. 2.3. Evaluating the importance of precursor material in the context of environmental friendly and cost effective carbon fibre production, increasing the tolerable heating rates or lowering the necessary stabilization temperature or time therefore could lower costs, overall processing times, energy demands and in turn environmental impact [16, 17].

As early as the 1950s and 1960s researchers tried to establish the reactions contributing to the change of chemical structure of PAN during stabilization. While significant progress has been made and principal reactions are widely accepted today, the full reaction mechanism is yet to be determined [18]. As the mechanisms of material transformation are not fully understood yet, there is a wide range of studies available trying to understand the processes and improve the precursor material. Al Aini et al. [19] undertook an extensive review of research of the last decades outlining the development that lead to today's commonly used precursor material and the current understanding of reaction mechanisms. In a study analysing the effect of microstructural orientation in PAN precursor material on the properties of carbon fibres, Wang et al. [20] showed in detail, that high orientation in fibres improved the cyclisation and dehydrogenation processes which in turn improved crosslinking reactions. Furthermore strength and modulus were found to be higher for more oriented material. Mittal et al. [21] reviewed fibre modification approaches after spinning, including coatings, treatment with catalysts and compounds and post-spinning stretching in various atmospheres. They found that there are a multitude of ways to improve the mechanical properties of the material within the mentioned categories. Bahl et al. [22] studied the effect of pre-treatment of the precursor material with Copper(I) chloride, reporting the treatment may help lower the activation energy and improve mechanical properties. Jie et al. [23] reported that the impregnation of PAN precursor with an aqueous $KMnO_4$ solution helped improve the density and orientation among other things leading to improved tensile strength. In another detailed study, itaconic acid as a comonomer was shown to lower the initiation temperature of the cyclization reaction while also broadening the exothermic peak, making the stabilization process better controllable [24]. Hao et al. [25] showed that the addition of methyl acrylate as a comonomer improves the reached cyclization degree leading to a better thermal stability and an overall higher carbon yield, an important factor from an environmental and cost perspective. Ammonium persulfate was used by Rahman et al. [26] to pre-treat PAN precursor in a study that concludes that the pre-treatment reduces necessary stabilization times and helps develop the cyclised structure. Goa et al. [27] added low amounts of graphene to the dope and observed that the spun fibres had improved mechanical properties and a lower void content, which resulted in carbon fibres with roughly 200% improved strength and modulus. Similarly Salim et al. [28] added graphene oxide to the precursor dope and also found that mechanical properties of resulting carbon fibres significantly increased. In an extensive review, Ahn et al. [29] showed the multitude of approaches explored to improve the precursor material towards high performance carbon fibres. They looked at efforts with new copolymers, increasing the molecular weight of the precursor, optimized dope compositions, spinning process optimization and development of new spinning processes. They conclude that that for any of the approaches to be successful, feasibility on industrial scale has to be demonstrated.

All shown approaches aim to improve the precursor material in a way that helps facilitate better processing in subsequent production steps. Some of the approaches are more tailored towards better material properties, while others focus on cost and processability. In most cases however additional research is necessary and is currently ongoing to prove and implement these ideas on a larger scale.

2.2.2 Textile PAN

Due to the high price of PAN typically used for carbon fibre production, other acrylic fibres commonly used for textile products, such as clothes or blankets, may be viable alternatives to reduce the overall cost of the final product. The term textile PAN, in the context of this study, refers to PAN-based material that promises improvements over traditionally used PAN precursor in terms of cost, energy and/or environmental impact. The main differences of these fibres compared to the special-grade fibres used in traditional carbon fibre production can be divided into two main categories, tow size and chemical composition, which may affect other properties of the fibres such as mechanical properties, diameter, density or molecular weight [30]. From a chemistry perspective textile grade PAN material generally contains at least 85% acrylonitrile, while PAN used for carbon fibre production typically consists of at least 90% acrylonitrile, the remainder of the material is made up of comonomers, 15 and 10% for textile and special grade material respectively [30, 31]. The added comonomers serve to improve processability and tune the material to the desired application. For textile materials dyes and other chemical substances may be present depending on the originally intended application. When carbonizing the material however, higher amounts of comonomers and impurities lead to increased mass loss and negatively affect the achievable mechanical properties [32]. Comonomers such as methyl acrylate or vinyl acetate are typically used to improve spinnability while acidic comonomers such as itaconic acid improve the processability during stabilization by broadening the exothermic peak [9, 33]. Industrial precursors typically combine two or more comonomers to improve several relevant process parameters [9].

To combine the strength of both types of material Khan et al. [32] investigated blending textile and special grade PAN with a ratio of 70/30% (textile/special grade) and found that the material could successfully be wet-spun, stabilized and subsequently carbonized, achieving mechanical properties of 2.2 GPa tensile strength and 210 GPa initial tensile modulus. Choi et al. [30] discussed a number of studies investigating different comonomer compositions and modified stabilization methods, such as electron beam radiation or plasma treatment. All studies showed to be cost-effective, but none are yet applicable on an industrial scale. Hiremath et al. [34] investigated a large tow size (450,000–600,000 filaments) textile grade precursor material of unspecified composition and found that mechanical properties achievable are in the range of commercially available moderate cost carbon fibre. Koenig et al. [35] report on the manufacture of carbon fibres from three types of textile grade polymers, two copolymers and one homopolymer. They used electron-beam radiation to lower the stabilization onset temperature and improve the microstructure of the material, consequently enhancing the achievable mechanical properties. Tensile strength reaches values of up to 3.1 ± 0.6 GPa and Young's moduli of up to 212 ± 9 GPa and are therefore in the range of commercially available carbon fibres. Jin et al. [56] investigated the structural evolution during the stabilization and carbonization process of a textile PNA precursor achieving a strength of 3.25 GPa and a modulus of 245 GPa. In a study comparing interfacial adhesion of three composites, Lee et al. [57] find that the strength and modulus of the carbon fibre composites made from special grade PAN and textile PAN exhibit very similar levels.

2.2.3 *Cellulose and Lignin*

Interestingly, cellulose is the oldest reported precursor material for carbon fibres, dating back to the late 1800s when the fibres were initially used in early versions of electric light bulbs [9]. It took however roughly another 100 years before the carbon fibre production process was more widely commercialised [10]. Both cellulose and lignin are more sustainable alternatives to the most commonly used PAN precursor material, with environmental as well as potential economic benefits. The cross-linked structure of cellulose, as shown in Fig. 2.3a, is beneficial for stabilization and carbonization reactions to progress uniformly, which helps achieve good material properties. Carbon fibres made from purified cellulose material typically have good mechanical properties, a high thermal conductivity and a high purity at relatively low cost [38, 39]. The biggest inhibitor of commercial success of the material as a carbon fibre precursor however is the low achievable yield of less than 30% [58]. Nevertheless research towards improving this type of precursor, especially the microstructure

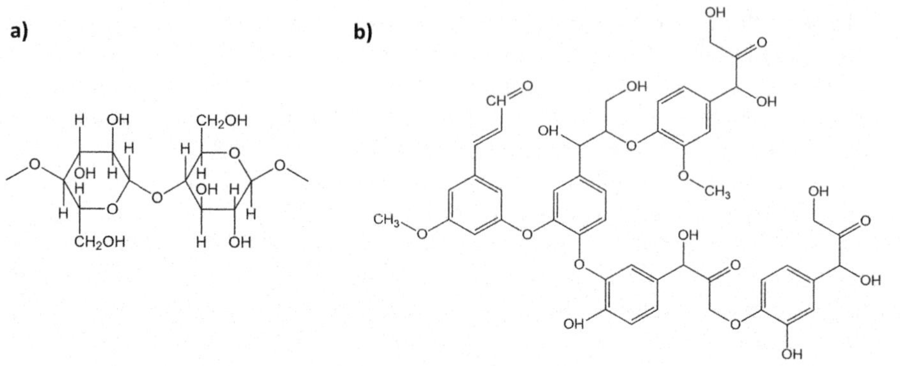

Fig. 2.3 Basic chemical structure of **a** cellulose and **b** lignin

and processing conditions, is ongoing. Furthermore interest is increasing in blending cellulose with other materials such as lignin [38, 39].

The use of lignin, a macromolecule present in about 1/3 of woody plants, may have the potential to save almost 50% of the production cost of carbon fibres [30, 40]. As lignin is a by-product of paper production, there is an abundant supply available at low cost. The high molecular carbon content makes it an ideal precursor source material. However, the unordered structure, shown in Fig. 2.3b, is detrimental to the processability of the material stabilization and carbonization conversion processes. The achievable yield, however, is significantly higher than cellulose and in the range of PAN with more than 50% [41]. Bajpai et al. [40] discussed in detail various types of lignin, i.e. from different plants, and possible production methods with different technologies, e.g. Kraft process, Steam Explosion, Organosolv, regarding processability and material properties. While publications show many different approaches and results vary, there is common ground when comparing to PAN-based carbon fibres. The tensile strength of lignin-based carbon fibres is much lower than that of PAN-based fibres (Lignin < 1000 MPa, PAN > 3000 MPa), while the average diameter is higher (Lignin > 10μm, PAN < 10μm) [61]. The properties of Lignin-based carbon fibres can greatly vary depending on the type and source, while for PAN-based carbon fibres the values are in a much smaller range, assuming that the carbon fibre production process has no influence [30, 40]. As for precursor fibre production, lignin is more versatile than PAN, as it can be wet spun, dry spun and melt spun with direct melt spinning being the most common method [40, 42]. As lignin is a natural material and typically only a by-product of other processes, e.g. paper-production, and therefore obtainable from a variety of sources, purity is another key factor for the processability of the material [63]. With improvement of the source material in terms of purity and material properties, lignin shows great potential as a precursor for carbon fibre production [40, 44].

In a review of approaches to improve known problems, Baker et al. [41] reported improvements of one of the following shortcomings: melt-spinning performance, fast carbon fibre conversion and cost efficiency. With a better understanding of the material properties on molecular level and detailed knowledge of necessary conversion processes, the potential of combining improvement strategies is promising. In another study Baker et al. [43] investigated the spinnability of types of hardwood obtained from black liquor as powder though kraft pulping and proprietary purification processes and concluded that only one type was successfully carbonizable, reaching modest material properties. Braun et al. [45] studied the stabilization process of purified hardwood kraft lignin outlining the occurring changes and concluded that for the used material an extremely low heating rate of 0.06 °C/min would be ideal. To combine good spinning performance with improved mechanical properties, Dai et al. [47] proposed an additional step between spinning and stabilization. Treating the freshly spun fibres with vaporized iodine was reported to raise the glass transition temperature of the material and improves the performance during stabilization/carbonization. The produced fibres showed better mechanical properties than untreated fibres. Choi et al. [30] reviewed several methods of source material purification published over the course of several years for different types and sources of lignin. Improved mechanical properties of the produced carbon fibre showed the success in optimizing the production process, as well as the source material properties. Furthermore the authors presented research on blending lignin with polymers, focusing on polyethylene, to further improve properties (tensile strength of up to 2.16 GPa and modulus of up to 130 GPa) and process capabilities, however with somewhat promising results [30]. Similarly Jing et al. [68] find that there is a limited range of PAN/lignin ratio in which carbon fibres with desirable material properties can be economically produced. For PAN with a lignin content higher than 20%, the mechanical properties of the subsequent carbon fibres worsen. The oxidation temperature however is positively influenced by higher lignin content, which would reduce energy usage/production cost.

One route to improving the shortcomings of the two green precursor materials that has been recently gaining traction, is a blending approach. Cellulose-based precursor fibre can be converted into high quality carbon fibre material with mechanical properties as high as 3.5 GPa tensile strengths and an initial tensile modulus of up to 690 GPa [9]. The main drawback of this material, however, is its low carbon fibre yield of around 10–30% owing to the relatively low initial carbon content of 44% [38, 48]. Lignin on the other hand has a significantly higher initial carbon content of over 60%, but due to the above shown complex and variable molecular structure, achievable mechanical properties are lower than for cellulose [49]. Trogen et al. [49] explored how different ratios of cellulose and lignin influence the achievable material properties, carbon fibre yield and processing behaviour, finding that for lignin concentrations of up to 30%, mechanical properties are only marginally impacted, but decrease significantly for higher ratios. Similarly Le et al. [50] investigated two cellulose-lignin blends (ratio 50/50 and 70/30) and how stabilization temperatures influence processing times and yield. In another study Le et al. [51] investigated the effect of boric acid treatment of the fibres on the process parameters and found that 3% boric acid led to a 25% reduction in stabilization time with increased thermal stability. The authors found that higher temperatures lead to reduced stabilization times, while higher lignin content increased the necessary time, but raises the overall achievable yield. Bengtsson et al. [52] studied stabilization conditions and precursor fibre diameters for a cellulose-lignin blend (30/70). They produced carbon fibres with an initial tensile modulus of 76 GPa and tensile strength of 1.07 GPa at a yield rate of 39%, however, at significantly longer stabilization times, than what is typical for industry standard PAN precursors. In another study Bengtsson et al. [48] investigated the carbon fibre yield for a cellulose-lignin blend (30/70). They were able to increase the yield from 22% for pure cellulose material to 40% for the blend while maintaining similar mechanical properties for the resulting carbon fibre. Comparing a continuous conversion process with a batch process to produce carbon fibre from a cellulose-lignin blend, Bengtsson et al. [53] also demonstrated that carbon fibres can be produced in a continuous process at similar processing times as current PAN processes, however with a lower yield of 30% and relatively low mechanical properties of up to 0.9 GPa strength and a modulus of 51 GPa.

Even though lignin/cellulose-based carbon fibres still have not reached the mechanical properties of PAN-based fibres due to issues in regard to source material type and quality, the potential of lignin-cellulose-blends as a precursor material remains considerable. One key aspect of carbon fibre made from lignin-cellulose-blends is the potentially lower price, due to cheaper source material and easier processability during precursor manufacturing, as the material can be melt-spun [41, 43]. This advantage however is counteracted by currently still significantly longer processing times for carbon fibre production or low material properties, as the above-mentioned current research results suggest. Stabilization heating rates vary between studies, starting as low as 0.01 °C/Min. [64] to 2–3 °C/Min. for purified, often pre-treated lignin-material [30, 40, 41, 54], up to 20–60 °C/Min. [41, 55] for extensively filtered/pre-treated fibres or blends. With low cost, the bio-based, fully renewable material still has the potential to become a great alternative for applications with higher monetary, less mechanical property focused requirements.

2.2.4 Alternative Source Materials

Besides improving or blending well established source materials to improve fibre properties or the production process, there is ongoing research effort towards introducing new materials to be converted into carbon fibres. Proposed materials include more traditional petroleum-based substances as well as fully renewable materials.

Due to a low price and wide availability polyethylene (PE) may be a viable substitute for PAN. As PE can be melt-spun, the costly solution spinning process can be avoided. Since PE degrades at high temperatures in an oxygen atmosphere, similar to PAN, but in this case undesired, the traditional stabilization process cannot be utilized. In a small-scale study, a thermochemical process with hot sulphuric acid was used instead. The carbonization stage corresponded to the PAN precursor process. PE-based carbon fibres showed a tensile strength of 1.9 GPa and a modulus of 178 GPa [56], which was lower than average values of PAN-based fibres. Nevertheless the mechanical properties reached levels required for use in automotive applications.

With a considerably lower projected price than PAN-based fibres (15 AUD vs. 31 AUD) for industrial scale production, PE may be a promising precursor alternative. Another potential precursor material investigated by Krumpfer et al. [57] is poly(methyl vinyl ketone). In a lab-scale process it was easily melt-spun, but had to be stored at low temperatures, in a dry place due to its low glass-transition temperature and the material absorbing water. The fibres could be stabilized in a chemical vapour-based process and had to be additionally heat-treated at 250 °C to not degrade during carbonization. While the research did not focus on the carbonization stage, initial studies showed promising results in terms of mechanical properties of the fibres. Further optimization of the process parameters was deemed necessary, but the approach overall proved poly(methyl vinyl ketone) may be a viable alternative precursor, enabling a less pollutant, less energy intensive production process. A more unconventional source material based on spider silk protein was investigated by Lewis et al. [58], where protein fibres were produced using transgenic silkworms, followed by several purification and preparation steps. After light-induced or chemical crosslinking, the fibres were then carbonized. The detailed process parameters still need to be investigated and optimized. The work showed an early-stage proof of concept with the goal of showing the possibility of a continuous production, besides general feasibility. Another approach focused on improving the most commonly used precursor PAN by sourcing it sustainably and cost efficiently. Towards this goal, Karp et al. [59] described a process converting ethyl 3-hydroxypropionic acid to acrylonitrile via alcohol dehydration and subsequent nitration. The ethyl 3-hydroxypropionic was derived microbially from lignocellulosic sugars. The price of the produced PAN was projected to be in the same range as petroleum-based PAN when produced at industrial scale. Sustainably sourced PAN would significantly improve greenhouse gas emissions of the carbon fibre product and may at larger scale also lower the price. Similarly Goyal et al. [60] proposed a process to convert glycerol or propylene glycol to PAN. Two novel catalytic processes were developed to archive the conversion by hydrocracking and dehydration. Through the subsequent ammoxidation bio-based acrylonitrile can be produced. While the process was sensitive to the purity of the used source material, fine-tuning the process parameters could improve the output quality. Furthermore the greenhouse gas emissions of the final product could be greatly improved [60]. Kenneth et al. [61] reviewed the current state of research to utilize more widely available and hence cheaper plastics, such as polyethylene, polyamide, polystyrene, polyester, and poly(vinyl chloride) for use as precursor, finding that polyethylene was more intensively investigated than other materials, but state that there may be considerable gain from investigating less prominent potential sources. Grasselli et al. [62] reviewed several approaches towards sustainably produced PAN, utilizing different source materials. In addition to the two concepts mentioned above, the production of acrylonitrile from glutamic acid was assessed as well as from oxygenated intermediates. Analysis of the glutamic process concludes that it was not feasible due to high cost and high environmental impact. Details on the oxygenated intermediates process are very scarce, but the concept itself was seen as promising, if it is cost-effective. Common ground for all processes was, that they were still in early-phase development and mostly not yet competitive with petroleum-based PAN. With ongoing research, rising awareness of the need for more sustainable products and increasing oil prices, there is high potential for bio-based PAN to become a competitive carbon fibre precursor candidate.

The majority of available data in any way linked to life cycle assessments, energy and material balances and cost models is based on processes using traditionally produced PAN as precursor. Data is available for lignin as the most prominent alternative sustainable precursor material, but comparability is low as some publications, for example include the production of the source material itself in greenhouse gas emissions calculation, while others consider lignin a by-product of paper-production without any (additional) climatic effects. For most of the alternative source materials mentioned in this review rough cost and emission estimations are made, but as most of the concepts are in early phases of development and mainly utilize research scale equipment, extensive life cycle assessments are not (yet) possible or feasible. An LCA/LCC model that can be easily adapted for different source materials and new technologies in an early research phase may have a beneficial impact on the development of new precursor materials, as an assessment in early development phases can hint at competitiveness of the material on a commercial scale and potential focus areas.

2.3 Carbon Fibre Manufacturing Process

A study of how to produce carbon fibres from PAN was first published in 1961 in Japan and the process was patented [9, 63]. The mechanical properties achievable using this precursor in combination with high yield led to increasing interest and adoption. The methods employed in carbon fibre production were concurrently refined in Japan, the United States and Great Britain [9, 10]. The basic manufacturing process can be divided into three major production steps, starting with the thermal stabilization of spun precursor, subsequent carbonization and surface treatment of carbonized fibre [10]. Figure 2.4 gives an overview of the process and its production steps.

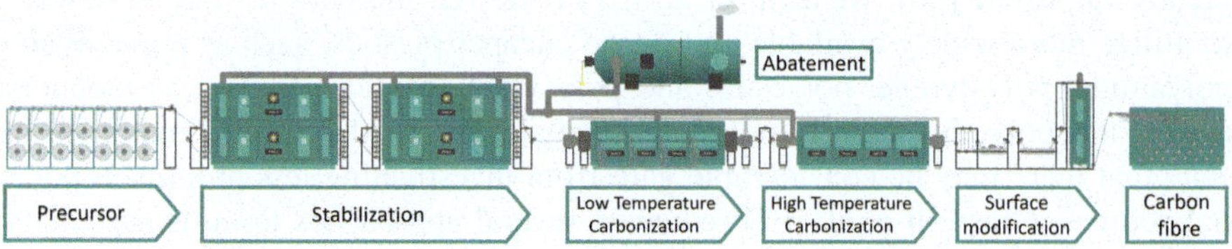

Fig. 2.4 Schematic representation of the carbon fibre manufacturing process

2.3.1 Process Setup and Requirements

The carbon fibre production process outlined here is based on PAN precursor, as it is currently the most commonly used precursor in industrial applications. The overall process however is similar for different commercial precursors currently in use.

Carbon fibres are generally defined to contain at least 92 wt% carbon [64, 65]. Their microstructure primarily consists of layered planar hexagonal networks of carbon atoms in a turbostratic arrangement consisting of crystalline and amorphous regions [65, 66]. Today, the vast majority of carbon fibre is manufactured from polyacrylonitrile (PAN) precursor fibres because of the excellent mechanical properties, such as a tensile strength as high as 7 GPa and modulus up to around 400 GPa, low density and excellent chemical stability, except in the presence of oxidizing agents at elevated temperatures [9, 64, 66, 67]. Fibre types are often categorised by their mechanical properties, such as high modulus (> 300 GPa), intermediate modulus (> 265–300 GPa), standard modulus (230–265 GPa) and high strength (> 3–4 GPa) [4, 65]. The excellent mechanical properties make the material particularly suitable for use in woven textile composites. Products or parts can then be manufactured through winding, moulding, vacuum bagging or other techniques [64].

The first manufacturing stage thermally stabilizes the fibres under tension in air at temperatures between 200 and 300 °C [14, 64] transforming the linear thermoplastic to a crosslinked polymer. The fibres pass through a sequence of ovens with increasing temperatures under controlled tension, gas flow and residence times, where the polyacrylonitrile chain structures are converted into a ladder type structure via cyclization, dehydrogenation and oxidation reactions [14, 68]. This process is a complex combination of chemical reactions and various models have been proposed to explain the conversion from PAN to oxidised PAN (OPF) fibre, such as those shown in Fig. 2.5 [64]. Quantifying the chemical transformations that occur during the carbon fibre manufacturing process, however, has proven to be difficult as the reactions are slow diffusion-driven processes. Additionally, multiple chemical reactions take place simultaneously and make it even more difficult to attribute changes in the material to specific reactions [69, 70].

The stabilization reactions and the rate at which they occur is critical to the final material quality achievable [10]. During this stage the density increases from around 1.18 g/cm^3 to around 1.37 g/cm^3. Cyclisation is one of the three key reactions, widely accepted, taking place during thermal stabilization [9, 10, 65]. Nitriles are converted into heterocyclic structures containing N in an exothermic reaction [9]. The second reaction taking place is dehydrogenation forming carbon–carbon double bonds via reaction of hydrogen from the polymer chain with oxygen from the oven atmosphere [71, 72]. The third reaction during the evolution of PAN into OPF is the oxidation. Oxygen has multiple effects during stabilization. The uptake of oxygen in the microstructure leads to better processability in the carbonization steps leading to improved final material and yield, and while it plays a part in initiating the cyclisation reaction it also raises the activation energy thus slowing the reaction progress [14, 71, 72].

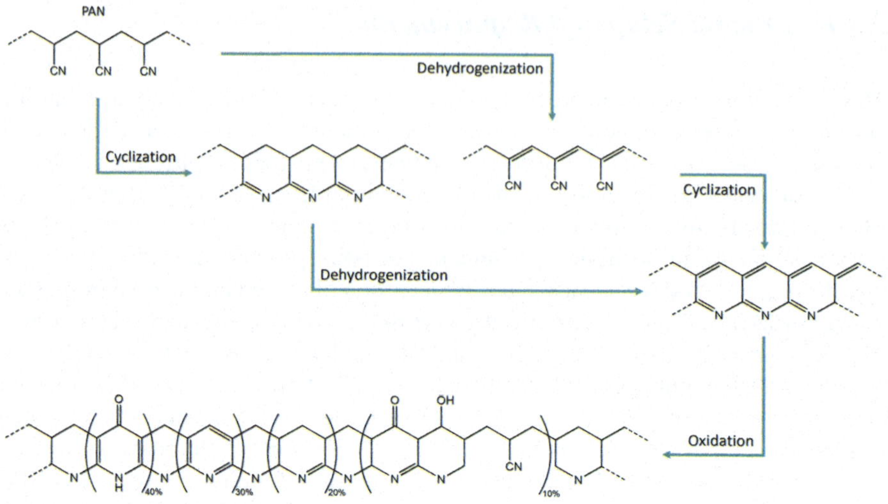

Fig. 2.5 Proposed chemical reactions during stabilization

Due to the exothermic nature of the stabilization process, the oven conditions are tightly controlled to avoid significant degradation and ensure a high-quality carbon fibre and yield [73, 74]. Reactive hazardous gases, such as CO, HCN and NH_3 evolve during this stage, requiring the heated air to be carefully circulated and partially exhausted to maintain emissions at concentrations which avoid explosive mixtures with air and maintain stable process conditions [9, 14, 64, 68].

After the stabilization stage, the fibres are heated in furnaces in only a few minutes from about 400 °C up to approximately 1600 °C during a two-step carbonization stage, low temperature (LT) and high temperature (HT) carbonization [9, 64]. The furnaces are typically purged with a constant flow of inert gas, such as nitrogen, to prevent oxygen from entering and to remove the evolved gaseous emissions. Up to approximately 600 °C a micro-structure with graphite-like ribbon structures is formed through dehydrogenation causing intermolecular cross-linking (Fig. 2.6) [14, 66]. At temperatures above 600 °C, denitrogenation causes these ribbons to further crosslink into sheet-like structures, creating the turbostratic micro-structure often referred to in literature [64, 66]. During the earlier stages of carbonization, up to 600 °C, H_2O, CO, CO_2 and NH_3, are primarily evolved. As the temperature increases, the composition of the emissions changes to include H_2, N_2 and HCN [9, 14]. During high temperature carbonization, further cross-linking occurs, the distance between the sheets is reduced, and the remaining nitrogen is liberated, leading to an increase in fibre modulus [9, 14, 66]. Finally, during the carbonization stage, fibre density increases and the fibre diameter is considerably reduced [14]. Specific temperature profile and the level of tension throughout the process affect the orientation of the fibre's internal formation and the size of the crystalline structures, which in turn determine the mechanical properties of the final fibres [14, 72, 75].

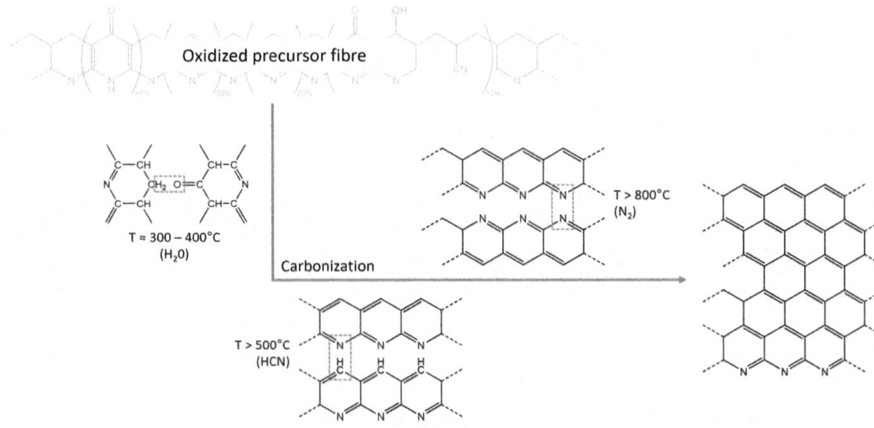

Fig. 2.6 Chemical transformation during the carbonization process

For high modulus carbon fibres, an additional graphitization step can be added to further improve the material. The treatment temperature is increased to up to 3000 °C [10, 14]. This step further increases the carbon content of the material and improves the microstructure, which increases the modulus of the fibres. The strength of the material however decreases, as voids are formed by the removal of remaining non-carbon elements and cross-links between the planar layers are reduced [76]. As the fibres contain only a very small amount of non-carbon elements at the beginning of this stage, only very little gas is evolved. Due to the high process temperatures, the use of extremely temperature tolerant equipment and use of argon to provide an inert atmosphere, graphitisation is a very expensive additional step only used for carbon fibres with very specific requirements [10, 14]. As this step is not used in the majority of applications, it is not further considered in this study. The LCA/LCC concept outlined in this work, however, can also be applied to assess graphitisation and may similarly yield insights into pathways to improve material cost and environmental impact.

The final process step before winding the material is dedicated to the modification and treatment of the carbon fibre surface to facilitate handling in downstream processes and increase fibre-matrix adhesion [14]. Typically, fibres are passed through an electrolytic bath containing an aqueous alkaline solution with the fibre material acting as the anode. In this process the fibre surface is chemically etched and functional groups introduced, enabling the interaction with resin matrix systems [9, 10]. The fibres are subsequently washed to remove any solution and the material is coated with a polymer-based sizing to improve processing in subsequent manufacturing steps and fibre-matrix bonding when incorporated in a composite material [9].

2.3.2 *Process Parameters, Their Influence and Other Considerations*

Process parameters are the key influence factor to adjust material properties through modifying the chemical and structural properties of the fibres. An in-depth understanding of how process parameters affect the chemical composition and microstructure of the fibre is critical to achieving a consistent material output with desired properties and a high yield rate. Fine tuning the process parameters not only can improve the produced material directly but also has an effect on the generated emissions and overall cost of the material. As simplified in Fig. 2.7, the relationship between process parameters, material characteristics and fibre properties can be considered circular in a research context. The process parameters define the material characteristics which in turn lead to certain fibre properties. These properties can then be used to inform decisions in regard to process parameter adjustments. While the parameters and properties shown in this simplified figure are critical, significantly more possible factors that characterise the carbon fibre manufacturing process may need to be considered, which is one of the reasons why commercial carbon fibre manufacturers carefully guard this information.

2.3.2.1 Stabilisation Process Parameters

In a continuous carbon fibre manufacturing process, during stabilisation the fibre material passes through a series of ovens with slowly increasing temperatures. A series of rollers at each oven exit continuously transport the fibre through multiple

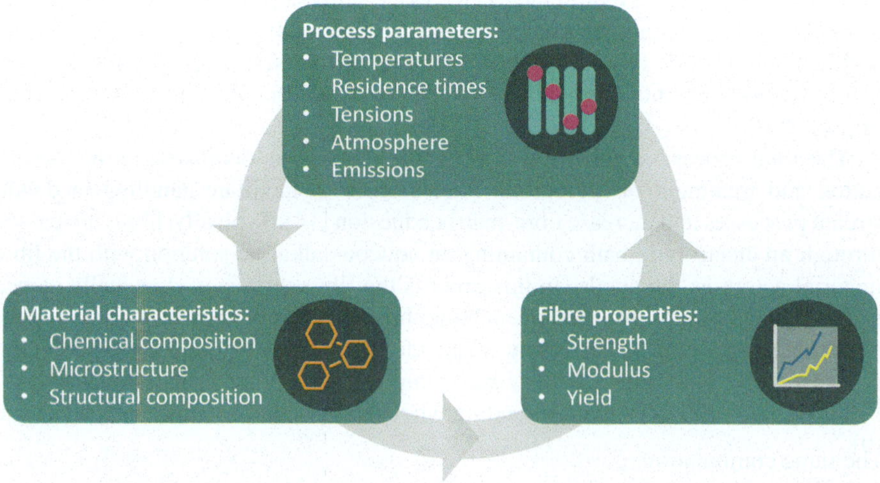

Fig. 2.7 Carbon fibre process—structure—property relationship

passes at the required speed. A larger number of ovens allows for faster processing but increases investment and space requirements [9, 77]. The ovens operate at normal atmospheric conditions as oxygen is needed for the occurring reactions to proceed [9]. As the chemical reactions during stabilisation are strongly exothermic, the process needs to be controlled tightly to allow ingress of enough air, remove sufficient amounts of generated emissions and not cause a thermal runaway, all while ensuring the desired properties are maintained and the speed is as high as possible for efficient manufacture. Due to the strong exothermic behaviour of PAN, tolerable heating rates are low, leading to long processing times of around 72 min. in heat [74]. Key parameters to control during the stabilization stage are the temperature profile, production line speed, tension, the atmospheric conditions inside the oven, i.e. the right amount of air ingress and emission removal, and others [9, 10, 14, 64, 68, 77, 78]. Due to the long residence time and large equipment size, stabilization is the largest contributor in terms of energy demand, especially when emission treatment costs are attributed on a production step level, reaching above 50% of the overall demand [67, 79, 80].

The temperature profile has to be chosen carefully, based on the used precursor and the desired material properties. In combination with the desired line speed, heating rates have to be kept below the limits of the precursor material. On the other hand, at heating rates and temperatures that are too low, the extent of reaction will not reach levels required to achieve the desired mechanical properties, or worse, the material will not withstand the high temperatures in the carbonization section. If the temperature profile or the heating rate is chosen too high, the material will be damaged or even a thermal runaway may occur with risk of fire and explosion [9, 14].

Tension is generated by drive systems, consisting of multiple rollers, before and after each process section running at slightly different speeds. The tension should be controlled carefully in each part of the process according to the occurring reactions and desired effect. The fibre shrinks due to stretched molecules relaxing at low temperatures and the cyclization reaction at mid-range temperatures. At higher temperatures within the stabilization process however, the sliding of polymer chains may cause elongation. These super-positioned effects entail meticulous individual adjustment of tensions for each sub-section [14].

The atmospheric conditions, especially the provision of sufficient oxidizing agent in the form of air, are critical to enable the earlier detailed chemical reactions to progress. The ideal oxygen content for best results was found to range between 15 and 21% [81]. As the fibres continuously move through the ovens, air is forcedly circulated to maintain an even temperature profile, constant atmospheric conditions and to remove generated emissions [9, 14].

2.3.2.2 Carbonization Process Parameters

The same combination of parameters is used to control the process in the carbonization stage, with a few differences. The atmosphere during carbonization is inert and therefore the furnaces need to be designed to avoid any air intake, while not excessively emitting inert gas at the furnace exits. Especially during low temperature carbonization, the exhaust flow needs to be kept high enough to remove the significant amount of evolved hazardous gases [9, 10, 14]. Tension plays a lesser role during carbonization as key carbon-containing structures have been already formed [14].

The above-mentioned process parameters contribute to the manufacture of high-quality carbon fibres, are strongly interconnected and therefore need to be well balanced to achieve good results. There is little information about processing conditions on industrial scale or even continuous scale, as these manufacturing specifications are closely kept secrets of the commercial carbon fibre manufacturers.

Besides the effect of the precursor material and the production parameters, other factors may affect the final materials properties. While there is research available regarding material storage and environmental temperature, humidity etc., focusing on the final composite products [82, 83], influence of these parameters on the production process, e.g. changing temperature or humidity during day- and night-shift production, and potential effects on the variability of the final products properties, are not well researched/documented. In fact, no publications on the topic could be found during this review of previous research efforts.

2.4 LCA for Carbon Fibre Production

Detailed assessments, covering the whole life cycle, are difficult to perform and require a high level of expert knowledge, especially for highly complex products like automobiles or airplanes. Therefore many approaches have been proposed to streamline assessments or improve the suitability for specific scenarios. This part introduces the basic concepts of LCA and covers more general research in the area.

2.4.1 LCA Methodology

LCA is a science-based multidisciplinary methodology to analyse environmental burdens linked to a product throughout its life cycle, starting with raw material extraction and ending with waste disposal [84, 85]. While the idea of assessing a products impact on and relationship with its environment was already developed in the 1960s, the Society of Environmental Toxicology and Chemistry (SETAC) was the driving force behind the first attempts to turn the existing concept into a more standardized methodology in the early 1990s [85–87]. This work became the basis for the first version of key international standards for LCA released towards the end of the 1990s [86]. With continuing development to more accurately measuring and attributing environmental interactions and expanding databases for these relations, these standards continue to be amended and improved. The methodology and general approach for LCA are standardized in ISO 14040:2006 and ISO 14044:2006 [3, 88].

These standards however only outline the very basic concept of the methodology, which leaves room for interpretation, adjustment to specific use-cases and improvement of the methodology itself. New approaches and methods, to cope with increasingly complex, global production systems, have been continuously presented and the development is still ongoing [85, 86]. Due to the complexity of the methodology stemming from the purposely incorporated flexibility and wide range of application, there are a number of international organizations attempting to increase transparency and international harmonious application. Among others, initiatives trying to foster better consensus are the Life Cycle Initiative of the United Nations Environment Program and the European Platform for LCA and the International Reference Life Cycle Data System [110]. The overall practical purpose of standardized LCAs can be summarized in four key points [3]:

- Improvement of the environmental performance of products throughout their life cycle or parts thereof
- Support for industrial, organizational or governmental decision-makers in fields such as strategy, planning, process design, prioritization or legislation
- Tracking of environmental performance with suitable indicators
- Product marketing and declaration.

This list is just a brief overview of key purposes. There are, however, many more scenarios where LCA can help to identify benefits or detriments of products [86].

LCA's typically consist of the four phases, goal and scope definition, inventory analysis, impact assessment and interpretation, shown in Fig. 2.8. They are depicted in a cycle and additionally interconnected as LCAs are often performed iteratively to improve accuracy [6].

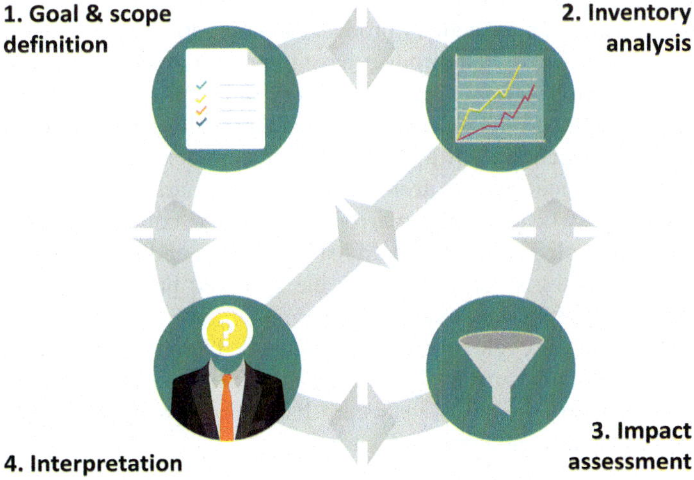

Fig. 2.8 Typical LCA phases

The first step of any assessment is the definition of the goal and scope of the study. The goal specifies the intended audience and outlines which questions should be answered and why they need to be answered. These definitions help to keep the study concise, select the appropriate methods and data sources in subsequent steps [3, 86, 88]. The scope of the study defines the system to be analysed, functions of the system to be included and where boundaries are drawn. As many products have a highly complex global production system and may be used around the world in different applications, it is important to limit the system boundaries of the study geographically and in terms of time [86, 90]. The boundary designation entails the definition of cut-off criteria, which should clearly outline which inputs and outputs are to be in- or excluded, based on their significance in terms of physical mass, energy and related environmental impact [88]. Due to size related rapidly increasing complexity, the system should generally be defined very carefully and kept as small as possible, while still including all relevant items contributing to a product's impact and ensuring that boundary crossing flows are either elementary or a product. The defined functional unit in essence quantitatively describes the product the assessment is performed for and serves as a reference for data normalization [86, 88]. Subsequently the reference flow has to be defined throughout the system and in mathematical correspondence with the defined functional unit [6, 88]. The methodology to assess impacts and types of relevant impacts should also be specified [88]. Finally the type of the study has to be defined, typically as either consequential, i.e. comparing scenarios how choosing alternatives impacts the studies outcomes or attributional, which evaluates the impacts of studied activities [86, 88]. The mentioned items are core elements to be considered, however thought should also be given to data requirements, data quality, potentially necessary assumptions, limitations and review processes [88].

The life cycle inventory (LCI) analysis constitutes the second phase of an LCA, where information and data about physical flows are collected. Any input of resources, materials, intermediary products and outputs such as emissions, waste, by-products and of course final product have to be accounted for [86, 88]. Natural laws such as the conservation of mass and energy can serve as a very basic guideline what to expect as a maximum in terms of in- and output, help avoid omitting relevant items and reduce inaccuracies [6]. Because of the complex nature of most analysed systems, the use of generic data from databases, summarized gate-to-gate data or estimated values may be necessary, where data cannot be collected directly [86, 90]. The resulting life cycle inventory should list all relevant and physically quantifiable flows in and out of the system related to the specified product and scaled to the amount linked to the functional unit [6, 86].

In the subsequent life cycle impact assessment (LCIA) phase, the collected data of the life cycle inventory is translated into environmental impacts based on standardized calculations and models. The ISO standard lists five steps for this phase, three of which are considered mandatory [3, 6, 86]:

1. *Selection* of impact categories that best support the assessment based on the goal and scope definition

2. In the *Classification* step the LCI results are assigned to impact categories selected in step 1. Every LCI item has to be assigned to all categories it contributes to.
3. The results of each category are then *characterized* based on defined environmental models generating a single numeric value with defined units for each category representing the overall environmental impact of the product in each category.

Optional steps:

4. *Normalization* can help increase the understanding of the generated results by relating them to common reference impacts
5. In a *Grouping* step impact categories are sorted and can be ranked for better comparison
6. *Weighting* may add a factor to each category to highlight and rank the perceived severity of different categories
7. *Data quality analysis* may clarify the reliability of the generated results, although this is in a similar form also part of the interpretation phase.

In the final interpretation phase, the generated results are summarized and interpreted to answer the initially proposed questions. Based on both the LCI and LCIA results, possible issues and limitations are presented. The interpretation should encompass all elements outlined in the goal and scope definition, while also be limited to the initially set boundaries [3, 6, 86, 88]. Sensitivity and uncertainty analysis provide information about the reliability of results and can hint at areas of focus for iterations or further work [86]. Within this phase a review and revising process may be implemented to improve quality and further increase reliability [6]. The methodology proposes an iterative approach to improve results and keep them up to date not only on the highest level, i.e. repeated full studies, but also suggests loops within and between the phases, whenever significant insights may lead to changes in other parts of the study [86].

Today's research in the field of LCA methodology often focuses on concepts to tailor the approach to specific use-cases or industries within the structure of the standards and guidelines. Another recurring topic is to integrate LCA in more holistic approaches including financial and social aspects on micro and macro scale.

This section provided a short overview of the development and basic structure of LCA limiting the excurse to what is most relevant to this work. Due to the high adaptability and wide-ranging application of the methodology, this could only be a superficial glimpse into the complex toolset that is Life Cycle Analysis.

2.4.2 Life Cycle Costing (LCC)

As carbon fibre is a high value product with significant cost associated, there is good reason to include a monetary component when assessing the material from a life cycle perspective. Live cycle costing is a reasonable choice, as it has a fair amount of overlap with the LCA methodology, specifically environmental LCC including cost of external factors such as environmental impacts.

LCC as a tool was in use as early as the 1930–60s in the US and gained traction well before sustainability and LCA. The realisation was that the initial purchase cost only represents a fraction of the overall cost of a product. In fact the cost of a product during the use-stage may be significantly higher, e.g. electric fan heater [91, 92]. With increasing awareness of resource scarcity and prices rising, LCC is becoming ever more widespread around the world [91, 93]. Despite this long history, unlike LCA there are still no cross-industrial international standards or widely adopted guidelines available detailing the methodology [93]. There are however sector-specific international standards, such as ISO 15686-5 [4], that additionally can serve as reference for the definitions used in this manuscript. LCC very basically can be defined as the cost of a product, encompassing everything from the initial research and design, manufacture, cost during operation until end of life [92]. Development and consensus is still building on methodological specifics; in recent years however three types have been classified within LCC: *conventional LCC*, which focuses on real costs, directly attributable to the product during the various stages, *environmental LCC*, which extends the scope to include more intangible and anticipated costs related to e.g. environmental impacts and finally *societal LCC*, which further extends the scope to also include external costs form a perspective of overall society [91]. Due to the lack of globally accepted standards, published models or analysis concepts tend to be very application specific and tailored to the needs of the applicant [94]. Woodward [92] describes the basic process of LCC based on a four phase model proposed by Harvey [95]. First all cash flows throughout the products life cycle and in any way directly related to it have to be identified. In a second step these costs have to be grouped into categories to be able to analyse trade-offs. The level of detail here is much dependent on the intention of the study. Subsequently mathematical descriptions to express the cost of individual items or steps have to be established, before finally an evaluation method is to be established. Based thereon and in conjunction with the LCA methodology, environmental LCC can be structured in the four phases, I. Goal and scope definition, II. Information analysis, III. Interpretation and hotspot analysis and IV. Discussion and sensitivity analysis [91]. Much like in an LCA, the goal and scope of the study need to be well defined initially with a functional unit, flows and system boundaries. In the subsequent phase all available cost data are gathered or substituted/estimated if not available. In a third step the gathered information is interpreted and impacts are analysed. Here hot spots can be identified. Finally the findings are discussed and conclusions can be drawn. Sensitivity analysis should be part of any environmental LCC to increase confidence in the drawn conclusions [91, 92].

Research related to carbon fibres in the field of LCC is relatively scarce and most recent publications focus on integrating the methodology with other approaches, often LCA. There are a few publications available with a focus on composites, only using aggregated database values for the manufacturing stage of the fibres, in various applications [96–101]. The closest available research based on actual gathered data rather than other publications or databases, was conducted by Nunna et al. [80]. They proposed a cost model for the manufacture of carbon fibre tows of various tow sizes including precursor manufacture and carbon fibre production. The model proposed, however, does not include any external costs such as environmental impacts.

2.4.3 LCA Holistic Concepts

In recent years LCA-centred research has often focused on broadening the assessment by combining methodologies and incorporating other types of analyses. With widespread adoption of LCA as a tool across industries, awareness has risen, that LCA with the inherent strong focus on environmental interactions cannot solely be used to inform decisions or fully compare products without trade-offs [86, 102]. From a broad sustainability perspective typically three dimensions have to be considered, financial, environmental and social [2]. Some of these aspects can be addressed within or by expanding the LCA methodology, e.g. focusing on the social aspect in social LCA [103]. The multidisciplinary nature of the LCA framework allows for wide ranging adaption and is open to incorporation of other methods and models [102].

For carbon fibre production Der et al. [104] proposed a multi-level concept to evaluate the production process from an LCA perspective and improve the reliability of the generated results. This framework intended to enable a holistic multi-disciplinary view of carbon fibre production in terms of the associated environmental impact. The interdependency of desired material properties and necessary manufacturing process parameters strongly affects LCA results and therefore often makes it difficult to compare results of different studies. The proposed concept tried to enable a lifecycle-based gate-to-gate assessment, where parameter changes or individual improvement measures are more quantifiable. After the correlations between material properties and production process parameters were evaluated, empirical data was generated to validate the observed relationships. The process was then modelled based on the generated data and identified relations. To lower the necessary effort and speed up the assessment, only critical production procedures (oxidation, LT/HT furnace) were modelled in detail. All simulated process steps were subsequently developed into a multi-scale production process simulation to model dynamic behaviour for energy demand and material flows. Finally in an interdisciplinary assessment, material scientists, production operators, customers and process developers can analyse potential scenarios and their effects on material properties and production processes.

Similarly Al-Lami et al. [105] proposed a framework approach for carbon fibre processing with a focus on cost. Eco-efficiency focused LCA guidelines were used as a basis and joined with related methods that are more cost-driven (LCCA—life cycle cost analysis) or product-oriented (BPR—business process reengineering). The LCA-phases proposed by ISO 14040 were adjusted to include methodology and goals from the mentioned methods. After goal and scope definition, the process was analysed on a low level of basic flows within the process steps and problems were defined in detail. In the next phase a virtual model was developed based on the concept derived from the results of the previous steps. The algorithms of the virtual model then computed the impact for each flow. The results can serve as a decision-making support for more sustainable development and introduction of measures which can be directly implemented. This framework, much like standard LCA, was designed to be an iterative approach with validation phases after major steps such as concept, virtual model and implementation.

Many of the more recent publications regarding LCA methodology for specific production types had a holistic approach in common. Like the concepts of [104, 105], a framework was defined and various methodologies were combined to fit the specific needs. For example, in a cost oriented industrial environment, eco-impact focused LCA principles might need to be coupled with more direct monetary approaches to foster application and acceptance. To this end, Abbate et al. [106] combined LCA and LCC to assess the viability of repaired, recycled and partly bio-based carbon fibre reinforced plastics. The two methodologies were applied in a concurrent fashion, suggesting that repair would be the most desirable solution. Similarly La Rose et al. [107] applied LCA and LCC methodology in a study examining chemical recycling processes of carbon fibre composites. Other studies also took a life cycle engineering perspective assessing performance, environmental impact and cost [108, 109]. In an extensive comparison of recent publications regarding circular economy performance assessments Sassanelli et al. [110] found, that many authors focused on a certain type of LCA variable, like material flows or energy flows, and therefore may disregard valuable information. The idea of the above-mentioned framework concepts was to try to shift this focused type of evaluation to more inclusive approaches.

2.5 Carbon Fibre Production Data-Based Studies

Although a number of publications discuss carbon fibre manufacture either in terms of life cycle analysis or the manufacturing process in a larger context, only very few studies provide detailed process specifics of the carbon fibre production process, as outlined below.

2.5.1 Carbon Fibre Production LCA Studies

There are only very few LCA studies publicly available specifically focused on carbon fibre production. Direct comparability of energy demands across the process and related results is limited as no standardized process step definitions exist. Hence, many publications only provide overall evaluation of major process steps or leave out relevant information about the production equipment. Nevertheless, these results can serve as reference points for current studies to some extent. Hermansson et al. [111] conducted a study comparing the environmental impact of carbon fibres and glass fibres in vehicle structures. While data for the carbon fibre production part of the assessment was collected, specifics are not mentioned due to confidentiality of the project. In another study, Hermannsson et al. [112] compare lignin-based and recycled carbon fibres in carbon fibre reinforced plastics (CFRP) with reported values for virgin material. They find that the information available is still quite limited for both materials and that uncertainties are relatively high.

Table 2.1 shows the total energy consumption per kilogram carbon fibre (MJ/kg) for the PAN-based carbon fibre manufacturing process from several published assessments highlighting the variability from study to study. If provided by the respective study, further details are outlined based on the three main production stages, stabilization, carbonization and surface treatment, which includes fibre sizing. Furthermore, the utilized fibre tow size and the overall process yield are given, as both may influence total energy demand.

The data of Arnold et al. [113] was one of the more comprehensive in terms of system description and energy and material flows, however, the energy demand values were reported by energy types, i.e. natural gas and electricity, rather than process steps. The calculations were based on a 12k tow line with an annual output of 1500t and a yield of 42%. Das [114] provided only very limited information about the observed production line, mentioning a 50k tow size and 48% yield, but concluded with the second highest total energy demand per kg carbon fibre. Suzuki et al. [115] did not report any other value than the total energy demand. Griffin et al. [116] did not mention a value for total energy demand as the research was focused on environmental impacts of the material flows. They concluded with the highest yield of 55% of all reviewed publications. A study conducted by Liddell et al. [117] for the US department of energy only showed average numbers for the US carbon fibre industry and furthermore only gave combined values for oxidation

Table 2.1 Reported energy consumption per kg carbon fibre

Source	Tow size	CF yield [%]	Stabilization [MJ/kg]	Carbonization [MJ/kg]	Surface treatment [MJ/kg]	Total (MJ/kg)
Arnold et al. [113]	12k	42	–	–	–	126
Das [114]	50k	48	–	–	–	459
Suzuki et al. [115]	–	–	–	–	–	286
Griffing [116]	–	55	–	4.47	0.05	–
Liddell et al./US DOE [117]	–	–	195		24	219
Liddell et al. [118]	–	–	175		20	195
Der et al. [119]	24k	44	166	258	–	1150

and carbonization. In another study Liddell et al. [118] used a similar approach, with slightly lower numbers for a traditional "state of the art" process. The values of Der et al. [119] were based on the Carbon Nexus 120t/year pilot production line and were calculated considering maintenance and peripheral processes, such as cooling/heating, lighting and compressed air supply. The overall yield achieved with this setup was 44%. Numerous other publications regarding LCA-data/energy flows for carbon fibre production lines referenced the mentioned publications, used database values or only gave estimations without further details and were within the range of energy demand summarised in the table above [7, 120–132].

Almost all of these studies lack comprehensiveness in terms of detailed data analysis and system description, making it difficult to compare the results, even if values for energy consumption or emissions are given. A standardized modular assessment approach, that can be adapted to various production scenarios, may streamline the LCA process and enable more comparable results for different production environments.

2.5.2 Studies Into Varying Production Parameters

While factors such as precursor material and properties may affect the production and properties of carbon fibres, the set production parameters mentioned above have significant influence on successful fibre production. Although parts of the reactions occurring, especially during stabilization, are still not fully understood, considerable research effort has been presented in the literature towards developing an in-depth understanding and enhancement of the process by modification of selected parameters.

Liu et al. [133] investigated the influence of air flow during oxidation using melt-spun PAN-precursor material, where the fibres were stabilized in a five-zone oven with temperatures ranging from 180 to 280 °C. Fibres produced at different airflow rates in the stabilization oven were then analysed. Results showed that the airflow did not have an effect on the chemical structure converted fibre. However, with increasing airflow the crystallinity and size of the crystals increased until a threshold was reached, above which carbon fibre yield started to decrease significantly. Nunna et al. [94] investigated the influence of stabilization dwell time and found that the density and crystal size of the oxidized precursor material can be adjusted through dwell time changes. Wu et al. [134] reported on the influence of tension during stabilization on the microstructure of the precursor material and suggest mechanisms causing the observed improvement in tensile strength. They suggested that molecular chains in amorphous regions in the fibre material were stretched first and in turn caused disorganized regions to be repacked into more ordered structures, while at higher temperatures stretching had a favourable effect on cyclization. Similar findings were reported by Wang et al. [135].

Wang et al. [136] investigated the microstructural evolution of oxidized precursor fibre in the low temperature carbonization outlining occurring reactions and likely mechanisms. Similarly Lee et al. [137] studied the microstructure changes throughout the carbonization of oxidised precursor fibre focusing on the radial evolution of changes. They confirmed the mechanisms widely reported in literature but also outlined differences between core and surface areas.

2.5.2.1 Optimisation Model Approaches

A stochastic optimization model developed by Khayyam et al. [138] focuses on the carbonization part of the production process. Using various stochastic methods, depending on the specific parameter (i.e. gas flow rate, drive speed) and the combination of the individual results, an energy balance model was deduced. The model was experimentally verified for various process parameter sets and desired material properties. Terra et al. [139] suggest a process optimization approach based on intelligent algorithms for key parameters of the stabilization process. Al Aini et al. [19] summarised new approaches that have been developed to decrease the necessary oxidation time and to increase the output.

Three methods mentioned in literature have been already licensed in the United States. The first, proposed by Paulauskas et al. [140], patented a concept where the stabilization process is enhanced by using an ozonous gas containing atmosphere at a controlled temperature. The second method more generally described a process where the PAN-fibres were stabilized and cross-linked in an aggressively oxidizing atmosphere using a highly reactive oxidizing species. As these fibres did not reach property levels of traditionally stabilized material, the field of application may be limited [141]. The third patent by Naskar et al. [142] proposed microwave assisted carbonization of stabilized PAN fibres. Liu et al. [143] similarly proposed the use of microwaves to enhance the production process. The approach however focused on the oxidation stage using microwave heating to stabilize the PAN-fibres. The process time as well as the necessary temperatures to produce high quality carbon fibres could be lowered. Elagib et al. [144] tried to enhance this concept by coating the precursor fibres with carbon black nanoparticles first, to improve absorbance of microwaves and optimize the (pre-) stabilization process. The produced pre-oxidized material showed favourable properties to produce high quality carbon fibres. Sunil et al. [145] reviewed the effect of surface treatment of the precursor fibres with different chemicals before thermal stabilization. Fibres pre-treated with guanidium carbonate or an ammonia solution accelerated the stabilization and therefore reduced the necessary process time. Dang et al. [146] proposed a heat treatment of PAN-fibres in nitrogen atmosphere prior to the stabilization stage. The treatment, only lasting in the range of hundreds of seconds, helped induce cyclization reactions. In turn the stabilization time could be reduced, leading to cost and time savings. Lu et al. [147] incorporated 10% carbon nanotubes (CNTs) in the PAN precursor and investigated the effect on processing time and material properties. The authors found that, while the incorporation of CNTs has both beneficial and detrimental effects on the microstructure evolution, the overall processing time can be reduced.

Zhao et al. [148] report on a method using microwave assisted low temperature carbonization, finding that compared to a traditional manufacturing process tensile strength and modulus could be slightly improved while running a 200 °C cooler process.

As shown above, there is ongoing research related to the carbon fibre manufacturing process with a variety of approaches and perspectives. Most studies however rely on small scale equipment not fully representative of an industrial manufacturing setup. The high investment and running cost of a scaled-up carbon fibre production facility makes them scarce around the world and therefore research on a larger scale is not widespread. The secrecy of commercial manufacturers further hinders a better understanding of the processes on a larger continuous scale, representative of industrial processing.

2.6 Circular Carbon Fibre Production

In times of climate change and increasing resource scarcity, the importance of circular production concepts, i.e. recycling, is becoming ever more important and research on the matter is being published more frequently. While for some materials, such as metals, recycling has been already a commonly adopted practice. For other materials, such as plastics, material recovery is not practiced, often for economic reasons [149–151]. Zink et al. [168] propose that the advantages of recycling can be summarized into one key benefit, displacement of primary production. This encompasses other benefits such as avoidance of final disposal and resource depletion and helps steer economies towards a more sustainable future [149, 152–154].

Material waste can be classified in various ways. One such approach is to relate generated waste to the life cycle stage it was produced in, leading to the terminology of post-industrial or pre-consumer waste for material incurred up until the use phase, and post-consumer waste for products and material discarded after their intended use [154, 155]. Carbon fibre waste categorized like this poses different challenges and requires different recycling strategies for pre- and post-consumer waste. While post-consumer waste typically comes in the form of composites with two or more distinct but defined materials present, waste incurred during carbon fibre production does often not have well defined material properties and due to the typically low quantities is also not well researched.

Another way of categorizing material waste streams is recycling in a closed loop, where the same products can be repeatedly made from the recycled material. Open loop recycling conversely may reuse the material for other products with lesser requirements in terms of material properties as employed recycling processes may impact achievable material properties [149, 154, 156]. Carbon fibres typically fall in the latter second-use type category, as cut carbon fibre fabrics cannot be returned to their original state of yarn on a spool [120, 150]. In this context it is also critical to evaluate the rate of recycling. While higher rates of material recovery can be desirable, the cost of achieving a 100% recycling rate may not be the most feasible way to increase acceptance.

2.6.1 Today's Carbon Fibre Recycling Concepts

Current carbon fibre recycling concepts mostly follow a second use concept rather than a fully circular approach. Furthermore the approaches usually focus on the recovery of either the fibres or the matrix material through mechanical, chemical or thermal processes. Pakdel et al. [120] reviewed current methods used for recycling carbon fibre composite materials. They showed the progress in the three main recycling routes currently employed, mechanical, thermal and chemical recycling, highlighting the effect of each type of approach on material properties. They concluded that thermal and mechanical processes were the more promising options. Howarth et al. [157] reviewed the energy intensity and environmental impact of mechanically recycled fibres. In the assessed process CFRP's manufactured from epoxy resin prepregs with a carbon fibre fraction of 55% were milled with various process parameters (milling speed, cutting tooth width etc.). The produced material can be used as particulate filler for example in coatings or SMC applications. Even though an industrial scale process would be energy intensive, it still showed a significant energy and environmental benefit, when compared to virgin carbon fibres. The second recycling concept used thermal processes to reclaim material. To optimize the commercially widespread pyrolysis process, Guo et al. [158] proposed to carry out the thermal decomposition under argon atmosphere. The fibres were pyrolyzed at up to 800 °C with a defined heating rate and then cut to be used in reinforced carbon–carbon composites. Although the surface of the produced fibres was slightly covered in residue from the pyrolytic reaction, the mechanical properties of the material were comparable to those of virgin carbon fibre. The fibres therefore could be used for the proposed purpose. ELG Carbon Fibre, a commercial carbon fibre recycler, similarly developed a modified pyrolysis process to reclaim high quality carbon fibres. The company produced a range of materials from the reclaimed fibres for automotive applications. The portfolio includes coatings and compounds from milled fibres, moulding compounds from chopped fibres and non-woven mats made from fibres up to 90mm long [62]. Chemical processes are another way of recycling carbon fibre composites. Okajima et al. [159] reviewed recycling concepts involving sub-/supercritical fluids (i.e. water, alcohol in different states/temperatures) with and without catalysts. While most assessed concepts produced fibres of close-to-virgin quality, there still are obstacles to commercialize these processes. Due to high temperatures and/ or pressures, cost and scalability are critical fields, which need research/ improvement. A similar method, proposed by Yu et al. [160], used ethylene glycol in combination with a covalent adaptable network epoxy resin to achieve a close to 100% recycling rate for both the fibres and the resin. While a high rate of recycling is of course desirable, it also has to be weighed against the effort to achieve such high rates and if there is a balance where the ratio of rate or recycling and effort are most economical and environmental-friendly. They stated, that reusing commonly employed thermosetting polymer matrix material was difficult due to its cross-linked network structure. The suggested covalent adaptable networks have similar properties in terms of thermal and chemical stability. However, using the right methods, the bonds in these networks could be broken and reconfigured and therefore the material could be fully recycled. The necessary reactions occur at temperatures well below the degradation temperature of the epoxy matrix and the carbon fibres. When the epoxy was fully dissolved, the carbon fibres could be cleanly separated, retaining properties like virgin fibres. The remaining solution then is further heated to evaporate the solvent and reclaim the polymer. Both the fibres and the polymer could then be used to mould new composite parts. Even though for high-tech use cases these materials still cannot replace virgin fibres and resin, many applications could benefit from this method, due to its environmental and monetary advantages. Arif et al. [161] reported on the recycling of glass and carbon-fibre composite materials for high value applications. They found that there were thermal and chemical recycling pathways to reuse the fibre material in high value applications, implementing such methods on industrial scale still seems challenging. Qazi et al. [162] propose a prediction model to optimize the thermal recycling process parameters for carbon fibres suggesting that use of such a model may help recover high value carbon fibre material for closed loop recycling.

While recycling remains an important area of research for carbon fibre-based products, as shown above, the overall impact of these processes must be carefully considered. While a common assumption is that it is desirable to recycle all material, an energy intensive recycling-process or highly hazardous by-products may cause overall greater damage than traditional end of life approaches like downcycling or landfilling.

2.6.2 Production Waste Management and Emission Treatment

With rising production volume of carbon fibres, the amount of waste generated during production is increasing. Similarly to recycling carbon fibre composite products, due to the high price of the source material and the final product, it is very important to find ways to reuse the incurred waste, besides generating as little waste as possible in the first place. However as the material transforms during the production process, it is more difficult to find simple recycling solutions without tailor-made approaches for each production step or fibre state. There are publications regarding the reuse of fully carbonized fibres that already have been mixed with resin [163, 164], but for waste material produced in the oxidation/carbonisation process, specifically during start up and power down, there is no research to be found. This may be due to the fact, that in relation to the output volume the amount of generated waste is relatively little and furthermore, as with other topics related to industrial carbon fibre, manufacturers do not publish any production related data. Nevertheless, there may be applications for this material that otherwise likely goes to landfill.

Another undesired output during the oxidation and carbonisation of the fibres are exhaust gases. The most commonly used precursor material, PAN, produces a variety of more or less toxic gases, which have to be treated before being released into the atmosphere. The most common process to treat the generated emissions is through thermal oxidation. There are several technical concepts achieving this, such as direct fired thermal oxidation or regenerative thermal oxidation and industry scale production facilities generally use different types of oxidation ovens and carbonization furnaces. The high-volume oxidation oven exhaust gas stream has a relatively low ration of toxic gases which need treatment. Therefore, regenerative thermal oxidation is the most efficient solution, as it needs little additional fuel and the heat can be easily recovered to increase efficiency. The exhaust gas stream generated in carbonization furnaces exits at a much lower flow rate but is heavily polluted. In order to efficiently remove contamination with low maintenance, direct incineration by using additional natural gas in a direct fired thermal oxidation system is often used. This much more energy-intensive process destructs pollutants at very high ratios of well over 90%. More recent developments try to integrate or combine technologies to further increase the overall efficiency and reduce pollution. Smaller (pilot-) production lines like the one found at Carbon Nexus, often only use one type of abatement system as the high cost of having more than one system only makes them feasible for high volume production [165–168]. What the mentioned types of abatement systems have in common is the fact that they only break down the most hazardous species of gases, but the overall amount of emissions increases further. From an environmental perspective they only reduce the immediate risk, but long-term effects related to global warming and similar phenomena are further increased. This also increases the burden on downstream processes and the use stage of carbon fibre products to offset these detrimental effects, when compared to other materials.

2.7 Review Summary and Research Gap

Various industries are seeking to lightweight/improve their products and are continuously exploring alternative materials to meet global emissions legislation. Carbon fibre-based material is a viable pathway to light-weighting in many cases, however, carbon fibre reinforced plastics (CFRP) are currently not cost competitive in high volume applications. The manufacture of the material requires significantly higher energy input compared to other lightweight engineering materials and currently most commonly used expensive commercial precursor material further increases the cost. Reducing the manufacturing costs and environmental impact by identification of energy saving hot-spots and process optimization potential in carbon fibre production will promote the penetration of carbon fibre into the new markets and improve competitiveness.

A combination of LCA and LCC can be used to explore the production environment from an environmental and economic perspective, e.g. investigating the energy and resource flows and identify non-value adding process activities. Systematically, energy saving hot-spots, emission heavy process steps and waste generating areas can be identified. While there is a lot of research available that addresses necessary steps to achieve improved properties of carbon fibres and/or the production process, most of the published work focuses on the fibre and on the production parameters. As outlined in the first part of this review, LCA and LCC relevant factors like energy demand or e.g. auxiliary material consumption are usually not considered in detail or only rough estimations are provided. Furthermore there is little high-level information available in terms of the influence of process parameter variation on these factors throughout the production line. LCAs conducted in the field of carbon fibre production can rarely be directly compared to one another as the scope varies and critical information is often missing. Additionally only a limited number of studies are publicly available as shown.

To better understand the effect and potential points of improvement for process parameters in terms of mass/energy balances, environmental and economic impacts, a holistic straightforward assessment approach for early phases of research is necessary. Through process analysis various production scenarios fostering energy efficient manufacture can be identified, experimentally validated and the results scaled. This will generate insight into opportunities to reduce the cost of the end product and can serve as a decision-making instrument for specifying efficient production settings.

Recent projects at Carbon Nexus have been dedicated to increasing the throughput of carbon fibre product by modifying processing parameters and the application of chemical-based acceleration processes in the carbon fibre manufacturing process. While the use of an industrial scale production line was unique to these works and made it possible to validate the research at large scale, the main focus of these projects was on parameter-structure–property relationships. The environmental and economic impacts were not studied in detail.

The aim of the proposed research is to assess and outline innovative, more energy efficient approaches to the manufacture of carbon fibre by the application of LCA and LCC. The project focuses on providing an in-depth understanding of the carbon fibre production from precursor to the final carbon fibre based on LCA and LCC principles. As there is no suitable methodological concept available which is tailored to the carbon fibre manufacturing process on a research scale, a concept to facilitate streamlined assessments of the manufacturing process without the need of specialized tools may yield quicker, more comparable results in early research phases. Using such a tool to explore the production environment from an environmental and economical perspective may yield new insights into pathways for improvement.

Gate-to-gate analyses of the carbon fibre production system, studying energy demands and resource flow, emissions and waste will help foster a better understanding of material-process-cost relationships. Varying processing conditions can help gain insights into environmental impact and cost drivers throughout the process. Analysing different precursor materials from a processing perspective, again with a focus on environmental and monetary impacts can steer research towards more promising materials. This project can help researchers better quantify economic and environmental impact in early phases of research.

The project will finally serve as a hub outlining the influence of product and process parameters on the cost and environmental impact of carbon fibre.

References

1. Hu Z (2022) A study of how soaring energy prices affect the economy. In: 2022-03-16T13:00:00.000Z 2022, Proceedings of the 7th international conference on economy, management, law and education (EMLE 2021): Atlantis Press, pp 78–83 [Online]. Available: https://doi.org/10.2991/aebmr.k.220306.012
2. Hauschild MZ, Herrmann C, Kara S (2017) An integrated framework for life cycle engineering. Procedia CIRP 61:2–9. https://doi.org/10.1016/j.procir.2016.11.257
3. ISO 14040 (2006) Environmental management—life cycle assessment—principles and framework, I. O. f. Standardization, Geneva, Switzerland [Online]. Available: https://books.google.com.au/books?id=C_xSygEACAAJ
4. ISO 15686-5:2017 (2017) Buildings and constructed assets—service life planning—Part 5: life-cycle costing I. O. f. Standardization, Geneva
5. Finkbeiner M, Schau EM, Lehmann A, Traverso M (2010) Towards life cycle sustainability assessment. Sustainability 2(10):3309–3322. https://doi.org/10.3390/su2103309
6. Klöpffer W, Grahl B, Klöpffer W (2014) Life Cycle Assessment (LCA): a guide to best practice. Wiley, Weinheim
7. Omnia LLC (2012) Assessment of carbon fiber manufacturing cost—extract from a benchmarking study on Iceland as a location for carbon fiber production. Omnia LLC, Iceland
8. Fry J (2016) The future of high capacity carbon fiber production. In: JEC World, Paris H. International (eds) Harper international, pp 1–37 [Online]. Available [Online]: http://www.harperintl.com/wp-content/uploads/2011/02/The-Future-of-High-Capacity-Carbon-Fiber-Production-Lines-Harper-International.pdf
9. Morgan P (2005) Carbon fibers and their composites. CRC Press, Ch. 1174, pp 185–259
10. Park S-J (2015) Carbon fibers (Springer Series in Materials Science). Springer Science+Business Media
11. Qin Y (2016) 3—A brief description of textile fibers. In: Qin Y (ed) Medical textile materials. Woodhead Publishing, pp 23–42
12. Kaur J, Millington K, Smith S (2016) Producing high-quality precursor polymer and fibers to achieve theoretical strength in carbon fibers: a review. J Appl Polym Sci 133(38):43963. https://doi.org/10.1002/app.43963
13. Rangarajan P et al (2002) Effect of comonomers on melt processability of polyacrylonitrile. J Appl Polym Sci 85(1):69–83. https://doi.org/10.1002/app.10655
14. Chung DDL (1994) Carbon fiber composites (no. 1). Heinemann, Newton, p 209
15. Morris EA, Weisenberger MC (2014) Solution spinning of PAN-based polymers for carbon fiber precursors. In: Polymer precursor-derived carbon, vol 1173, (ACS Symposium Series, no. 1173). American Chemical Society, Ch. 9, pp 189–213
16. Mathur RB, Bahl OP, Sivaram P (1992) Thermal degradation of polyacrylonitrile fibres. Curr Sci 62(10):662–669
17. Gupta AK, Paliwal DK, Bajaj P (1998) Melting behavior of acrylonitrile polymers. J Appl Polym Sci 70(13):2703–2709. https://doi.org/10.1002/(sici)1097-4628(19981226)70:13%3c2703::Aid-app15%3e3.0.Co;2-2
18. Bashir Z (1991) A critical review of the stabilisation of polyacrylonitrile. Carbon 29(8):1081–1090. https://doi.org/10.1016/0008-6223(91)90024-d

19. Al Aiti M, Jehnichen D, Fischer D, Brünig H, Heinrich G (2018) On the morphology and structure formation of carbon fibers from polymer precursor systems. Progr Mater Sci 98:477–551. https://doi.org/10.1016/j.pmatsci.2018.07.004
20. Wang B, Li C, Cao W (2021) Effect of polyacrylonitrile precursor orientation on the structures and properties of thermally stabilized carbon fiber. Materials (Basel) 14(12). https://doi.org/10.3390/ma14123237
21. Mittal J, Mathur RB, Bahl OP (1997) Post spinning modification of PAN fibres—a review. Carbon 35(12):1713–1721. https://doi.org/10.1016/s0008-6223(97)00126-7
22. Bahl DP, Mathur RB, Dhami TL (1985) Modification of polyacrylonitrile fibres to make them suitable for conversion into high performance carbon fibres. Mater Sci Eng 73:105–112. https://doi.org/10.1016/0025-5416(85)90300-3
23. Jie L, Wangxi Z (2005) Structural changes during the thermal stabilization of modified and original polyacrylonitrile precursors. J Appl Polym Sci 97(5):2047–2053. https://doi.org/10.1002/app.21916
24. Ouyang Q, Cheng L, Wang H, Li K (2008) Mechanism and kinetics of the stabilization reactions of itaconic acid-modified polyacrylonitrile. Polym Degrad Stab 93(8):1415–1421. https://doi.org/10.1016/j.polymdegradstab.2008.05.021
25. Hao J, Liu Y, Lu C (2018) Effect of acrylonitrile sequence distribution on the thermal stabilization reactions and carbon yields of poly(acrylonitrile-co-methyl acrylate). Polym Degrad Stab 147:89–96. https://doi.org/10.1016/j.polymdegradstab.2017.11.010
26. Rahman MM, Demirel T, Tunçel KŞ, Karacan I (2021) The effect of the ammonium persulfate and a multi-step annealing approach during thermal stabilization of polyacrylonitrile multifilament prior to carbonization. J Mater Sci 56(26):14844–14865. https://doi.org/10.1007/s10853-021-06209-1
27. Gao Z et al (2020) Graphene reinforced carbon fibers. Sci Adv 6(17):eaaz4191. https://doi.org/10.1126/sciadv.aaz4191
28. Salim NV, Jin X, Razal JM (2019) Polyacrylonitrile/liquid crystalline graphene oxide composite fibers—towards high performance carbon fiber precursors. Compos Sci Technol 182. https://doi.org/10.1016/j.compscitech.2019.107781
29. Ahn H, Yeo SY, Lee BS (2021) Designing materials and processes for strong polyacrylonitrile precursor fibers. Polymers (Basel) 13(17). https://doi.org/10.3390/polym13172863
30. Choi D, Kil H-S, Lee S (2019) Fabrication of low-cost carbon fibers using economical precursors and advanced processing technologies. Carbon 142:610–649. https://doi.org/10.1016/j.carbon.2018.10.028
31. Yoo SH et al (2017) Facile method to fabricate carbon fibers from textile-grade polyacrylonitrile fibers based on electron-beam irradiation and its effect on the subsequent thermal stabilization process. Carbon 118:106–113. https://doi.org/10.1016/j.carbon.2017.03.039
32. Khan H, Kaur J, Naebe M, Hutchinson S, Varley RJ (2022) Continuous, pilot-scale production of carbon fiber from a textile grade PAN polymer. Mater Today Commun 31. https://doi.org/10.1016/j.mtcomm.2022.103231
33. Tsai J-S, Lin C-H (1991) Effect of comonomer composition on the properties of polyacrylonitrile precursor and resulting carbon fiber. J Appl Polym Sci 43(4):679–685. https://doi.org/10.1002/app.1991.070430405
34. Hiremath N, Young S, Ghossein H, Penumadu D, Vaidya U, Theodore M (2020) Low cost textile-grade carbon-fiber epoxy composites for automotive and wind energy applications. Compos Part B: Eng 198. https://doi.org/10.1016/j.compositesb.2020.108156
35. König S et al (2021) High-performance carbon fibers prepared by continuous stabilization and carbonization of electron beam-irradiated textile grade polyacrylonitrile fibers. Macromolecul Mater Eng 306(12). https://doi.org/10.1002/mame.202100484
36. Jin X, Feng C, Creighton C, Hameed N, Parameswaranpillai J, Salim NV (2021) On the structural evolution of textile grade polyacrylonitrile fibers during stabilization and carbonization: towards the manufacture of low-cost carbon fiber. Polym Degradat Stabil 186. https://doi.org/10.1016/j.polymdegradstab.2021.109536

37. Lee S, Ham S, Youn SJ, Chung YS, Lee S (2021) Effect of textile PAN-based carbon fibers with rough surface on interfacial adhesion in PA6 composites. Ind Eng Chem Res 60(25):9088–9097. https://doi.org/10.1021/acs.iecr.1c01015
38. Dumanlı AG, Windle AH (2012) Carbon fibres from cellulosic precursors: a review. J Mater Sci 47(10):4236–4250. https://doi.org/10.1007/s10853-011-6081-8
39. Kadla JF, Kubo S, Venditti RA, Gilbert RD, Compere AL, Griffith W (2002) Lignin-based carbon fibers for composite fiber applications. Carbon 40(15):2913–2920. https://doi.org/10.1016/s0008-6223(02)00248-8
40. Bajpai P (2017) Lignin as a raw material for carbon fibre. In: Carbon fibre from lignin. SpringerBriefs in Materials: Springer, Chapter 5, pp 29–32
41. Baker DA, Rials TG (2013) Recent advances in low-cost carbon fiber manufacture from lignin. J Appl Polym Sci 130(2):713–728. https://doi.org/10.1002/app.39273
42. Wei J, Oksman K (2022) Manufacture and application of lignin-based carbon fibers and lignin-based carbon nanofibers. In: Micro and nanolignin in aqueous dispersions and polymers, pp 203–236
43. Baker DA, Gallego NC, Baker FS (2012) On the characterization and spinning of an organic-purified lignin toward the manufacture of low-cost carbon fiber. J Appl Polym Sci 124(1):227–234. https://doi.org/10.1002/app.33596
44. Li Z, Rosli RAB, Xiang Y, Ye E, Loh XJ (2018) Carbon precursor from lignin: methods and applications. In: Functional materials from lignin. Sustainable chemistry series, pp 121–152
45. Braun JL, Holtman KM, Kadla JF (2005) Lignin-based carbon fibers: oxidative thermostabilization of kraft lignin. Carbon 43(2):385–394. https://doi.org/10.1016/j.carbon.2004.09.027
46. Dai Z, Shi X, Liu H, Li H, Han Y, Zhou J (2018) High-strength lignin-based carbon fibers via a low-energy method. RSC Adv 8(3):1218–1224. https://doi.org/10.1039/c7ra10821d
47. Jiang X et al (2018) Preparation of low-cost carbon fiber precursors from blends of wheat straw lignin and commercial textile-grade polyacrylonitrile (PAN). Holzforschung 72(9):727–734. https://doi.org/10.1515/hf-2017-0191
48. Bengtsson A et al (2018) Improved yield of carbon fibres from cellulose and kraft lignin. Holzforschung 72(12):1007–1016. https://doi.org/10.1515/hf-2018-0028
49. Trogen M et al (2021) Cellulose-lignin composite fibres as precursors for carbon fibres. Part 1—manufacturing and properties of precursor fibres. Carbohydr Polym 252:117133. https://doi.org/10.1016/j.carbpol.2020.117133
50. Le N-D, Trogen M, Ma Y, Varley RJ, Hummel M, Byrne N (2020) Understanding the influence of key parameters on the stabilisation of cellulose-lignin composite fibres. Cellulose 28(2):911–919. https://doi.org/10.1007/s10570-020-03583-y
51. Le N-D, Trogen M, Varley RJ, Hummel M, Byrne N (2020) Effect of boric acid on the stabilisation of cellulose-lignin filaments as precursors for carbon fibres. Cellulose 28(2):729–739. https://doi.org/10.1007/s10570-020-03584-x
52. Bengtsson A, Bengtsson J, Sedin M, Sjöholm E (2019) Carbon fibers from lignin-cellulose precursors: effect of stabilization conditions. ACS Sustain Chem Eng 7(9):8440–8448. https://doi.org/10.1021/acssuschemeng.9b00108
53. Bengtsson A et al (2022) Continuous stabilization and carbonization of a lignin-cellulose precursor to carbon fiber. ACS Omega 7(19):16793–16802. https://doi.org/10.1021/acsomega.2c01806
54. Uraki Y, Nakatani A, Kubo S, Sano Y (2001) Preparation of activated carbon fibers with large specific surface area from softwood acetic acid lignin. J Wood Sci 47(6):465–469. https://doi.org/10.1007/bf00767899
55. Sudo K, Shimizu K, Nakashima N, Yokoyama A (1993) A new modification method of exploded lignin for the preparation of a carbon fiber precursor. J Appl Polym Sci 48(8):1485–1491. https://doi.org/10.1002/app.1993.070480817
56. De Palmenaer A, Wortberg G, Merke M, Roeding T, Gries T, Seide G (2018) Carbon fibre development: from polyethylene-based precursors to carbon fibres. In: Man made fiber year-book, pp 39–40

57. Krumpfer JW et al (2016) Poly(Methyl Vinyl Ketone) as a potential carbon fiber precursor. Chem Mater 29(2):780–788. https://doi.org/10.1021/acs.chemmater.6b04774
58. Lewis RV, Paulauskas FL, Ozcan S, Hayashi C (2017) Spider silk MASP1 and MASP2 proteins as carbon fiber precursors. Utah State University
59. Karp EM et al (2017) Renewable acrylonitrile production. Science 358(6368):1307–1310. https://doi.org/10.1126/science.aan1059
60. WBS 2.3.1.200 (2017) Biomass conversion to Acrylonitrile monomer-precursor for the production of carbon fibers
61. Brown KR, Harrell TM, Skrzypczak L, Scherschel A, Wu HF, Li X (2022) Carbon fibers derived from commodity polymers: a review. Carbon 196:422–439. https://doi.org/10.1016/j.carbon.2022.05.005
62. Grasselli RK, Trifirò F (2016) Acrylonitrile from biomass: still far from being a sustainable process. Top Catal 59(17–18):1651–1658. https://doi.org/10.1007/s11244-016-0679-7
63. Nakamura O et al (2009) Study on the PAN carbon-fiber-innovation for modeling a successful R&D management. Synthesiol Eng Edn 2(2):154–164. https://doi.org/10.5571/syntheng.2.154
64. Frank E, Hermanutz F, Buchmeiser MR (2012) Carbon fibers: precursors, manufacturing, and properties. Macromol Mater Eng 297(6):493–501. https://doi.org/10.1002/mame.201100406
65. Huang X (2009) Fabrication and properties of carbon fibers. Materials 2(4):2369–2403. https://doi.org/10.3390/ma2042369
66. Park S-J, Heo G-Y (2015) Precursors and manufacturing of carbon fibers. In: Park S-J (ed) Carbon fibers. Springer Series in Materials Science. Springer Netherlands, Dordrecht, Chapter 2, pp 31–66
67. Maghe M et al (2016) Using ionic liquids to reduce energy consumption for carbon fibre production. Journal of Materials Chemistry A 4(42):16619–16626. https://doi.org/10.1039/c6ta06842a
68. Fitzer E, Frohs W, Heine M (1986) Optimization of stabilization and carbonization treatment of PAN fibres and structural characterization of the resulting carbon fibres. Carbon 24(4):387–395. https://doi.org/10.1016/0008-6223(86)90257-5
69. Edie DD (1998) The effect of processing on the structure and properties of carbon fibers. Carbon 36(4):345–362. https://doi.org/10.1016/s0008-6223(97)00185-1
70. Nunna S, Naebe M, Hameed N, Fox BL, Creighton C (2017) Evolution of radial heterogeneity in polyacrylonitrile fibres during thermal stabilization: an overview. Polym Degrad Stab 136:20–30. https://doi.org/10.1016/j.polymdegradstab.2016.12.007
71. Fitzer E, Müller DJ (1975) The influence of oxygen on the chemical reactions during stabilization of pan as carbon fiber precursor. Carbon 13(1):63–69. https://doi.org/10.1016/0008-6223(75)90259-6
72. Rahaman MSA, Ismail AF, Mustafa A (2007) A review of heat treatment on polyacrylonitrile fiber. Polym Degrad Stab 92(8):1421–1432. https://doi.org/10.1016/j.polymdegradstab.2007.03.023
73. Nunna S, Maghe M, Fakhrhoseini S, Polisetti B, Naebe M (2018) A Pathway to reduce energy consumption in the thermal stabilization process of carbon fiber production. Energies 11(5):1145. https://doi.org/10.3390/en11051145
74. Nunna S et al (2019) Time dependent structure and property evolution in fibres during continuous carbon fibre manufacturing. Materials (Basel) 12(7):1069. https://doi.org/10.3390/ma12071069
75. Watt W (1997) Production and properties of high modulus carbon fibres. Proc R Soc Lond. A. Math Phys Sci 319(1536):5–15. https://doi.org/10.1098/rspa.1970.0161
76. Qin X, Lu Y, Xiao H, Wen Y, Yu T (2012) A comparison of the effect of graphitization on microstructures and properties of polyacrylonitrile and mesophase pitch-based carbon fibers. Carbon 50(12):4459–4469. https://doi.org/10.1016/j.carbon.2012.05.024
77. Frank E, Steudle LM, Ingildeev D, Sporl JM, Buchmeiser MR (2014) Carbon fibers: precursor systems, processing, structure, and properties. Angew Chem Int Ed 53(21):5262–5298. https://doi.org/10.1002/anie.201306129

78. Bajaj P, Roopanwal AK (1997) Thermal stabilization of acrylic precursors for the production of carbon fibers: an overview. J Macromolecul Sci Part C: Polym Rev 37(1):97–147. https://doi.org/10.1080/15321799708014734
79. International H (2013) Processing advancements within reach for achieving significant reductions in carbon fiber cost of manufacturing. In: JEC Europe 2013, ICS carbon conference, Paris. Harper International, Paris, p 41 [Online]. Available: http://www.harperintl.com/wp-content/uploads/2011/09/JEC.pdf
80. Nunna S, Blanchard P, Buckmaster D, Davis S, Naebe M (2019) Development of a cost model for the production of carbon fibres. Heliyon 5(10):e02698. https://doi.org/10.1016/j.heliyon.2019.e02698
81. Wang J, Hu L, Yang C, Zhao W, Lu Y (2016) Effects of oxygen content in the atmosphere on thermal oxidative stabilization of polyacrylonitrile fibers. RSC Adv 6(77):73404–73411. https://doi.org/10.1039/c6ra15308a
82. Dhieb H, Buijnsters JG, Elleuch K, Celis JP (2016) Effect of relative humidity and full immersion in water on friction, wear and debonding of unidirectional carbon fiber reinforced epoxy under reciprocating sliding. Compos B Eng 88:240–252. https://doi.org/10.1016/j.compositesb.2015.11.011
83. Wang S, Kowalik DP, Chung DDL (2010) Effects of the temperature, humidity, and stress on the interlaminar interface of carbon fiber polymer-matrix composites, studied by contact electrical resistivity measurement. J Adhes 78(2):189–200. https://doi.org/10.1080/00218460210384
84. Peña C et al (2021) Using life cycle assessment to achieve a circular economy. Int J Life Cycle Assess 26(2):215–220. https://doi.org/10.1007/s11367-020-01856-z
85. Klopffer W (1997) Life cycle assessment: from the beginning to the current state. Environ Sci Pollut Res Int 4(4):223–228. https://doi.org/10.1007/BF02986351
86. Hauschild MZ, Rosenbaum RK, Olsen SI (2018) Life cycle assessment, 1st edn. Springer International Publishing, Cham
87. Ayres RU (1995) Life cycle analysis: a critique. Resour Conserv Recycl 14(3–4):199–223. https://doi.org/10.1016/0921-3449(95)00017-d
88. ISO 14044 (2006) Environmental management—life cycle assessment—requirements and guidelines, I. O. f. Standardization, Geneva [Online]. Available: https://books.google.com.au/books?id=1SEkygAACAAJ
89. Finnveden G et al (2009) Recent developments in life cycle assessment. J Environ Manage 91(1):1–21. https://doi.org/10.1016/j.jenvman.2009.06.018
90. General guide for Life Cycle Assessment (2010) Provisions and action steps. Publications Office
91. Hunkeler D, Lichtenvort K, Rebitzer G (2008) Environmental life cycle costing. CRC Press
92. Woodward DG (1997) Life cycle costing—theory, information acquisition and application. Int J Project Manage 15(6):335–344. https://doi.org/10.1016/s0263-7863(96)00089-0
93. Korpi E, Ala-Risku T (2008) Life cycle costing: a review of published case studies. Manag Audit J 23(3):240–261. https://doi.org/10.1108/02686900810857703
94. Sherif YS, Kolarik WJ (1981) Life cycle costing: concept and practice. Omega 9(3):287–296. https://doi.org/10.1016/0305-0483(81)90035-9
95. Harvey G (1976) Life-cycle costing: a review of the technique. Manag Acc 57:343–347
96. Calado EA, Leite M, Silva A (2019) Integrating life cycle assessment (LCA) and life cycle costing (LCC) in the early phases of aircraft structural design: an elevator case study. Int J Life Cycle Assess 24(12):2091–2110. https://doi.org/10.1007/s11367-019-01632-8
97. Witik RA, Payet J, Michaud V, Ludwig C, Månson J-AE (2011) Assessing the life cycle costs and environmental performance of lightweight materials in automobile applications. Compos A Appl Sci Manuf 42(11):1694–1709. https://doi.org/10.1016/j.compositesa.2011.07.024
98. Bianchi I, Forcellese A, Marconi M, Simoncini M, Vita A, Castorani V (2021) Environmental impact assessment of zero waste approach for carbon fiber prepreg scraps. Sustain Mater Technol 29. https://doi.org/10.1016/j.susmat.2021.e00308

99. Tapper RJ, Longana ML, Norton A, Potter KD, Hamerton I (2020) An evaluation of life cycle assessment and its application to the closed-loop recycling of carbon fibre reinforced polymers. Compos B Eng 184:107665. https://doi.org/10.1016/j.compositesb.2019.107665
100. Meng F, Cui Y, Pickering S, McKechnie J (2020) From aviation to aviation: environmental and financial viability of closed-loop recycling of carbon fibre composite. Compos Part B: Eng 200. https://doi.org/10.1016/j.compositesb.2020.108362
101. Wheatley A, Warren D, Das S (2013) Low-cost carbon fibre: applications, performance and cost models. In: Advanced composite materials for automotive applications, pp 405–434
102. Zamagni A (2012) Life cycle sustainability assessment. Int J Life Cycle Assess 17(4):373–376. https://doi.org/10.1007/s11367-012-0389-8
103. Jørgensen A (2012) Social LCA—a way ahead? Int J Life Cycle Assess 18(2):296–299. https://doi.org/10.1007/s11367-012-0517-5
104. Dér A, Kaluza A, Kurle D, Herrmann C, Kara S, Varley R (2018) Life cycle engineering of carbon fibres for lightweight structures. Procedia CIRP 69:43–48. https://doi.org/10.1016/j.procir.2017.11.007
105. Al-Lami A, Hilmer P, Sinapius M (2018) Eco-efficiency assessment of manufacturing carbon fiber reinforced polymers (CFRP) in aerospace industry. Aerosp Sci Technol 79:669–678. https://doi.org/10.1016/j.ast.2018.06.020
106. Abbate E, Mirpourian M, Brondi C, Ballarino A, Copani G (2022)Environmental and economic assessment of repairable carbon-fiber-reinforced polymers in circular economy perspective. Materials (Basel) 15(9). https://doi.org/10.3390/ma15092986
107. La Rosa AD, Greco S, Tosto C, Cicala G (2021) LCA and LCC of a chemical recycling process of waste CF-thermoset composites for the production of novel CF-thermoplastic composites. Open loop and closed loop scenarios. J Clean Prod 304. https://doi.org/10.1016/j.jclepro.2021.127158
108. Carvalho H et al (2016) Application of life cycle engineering approach to assess the pertinence of using natural fibers in composites—the rocker case study. Procedia CIRP 48:364–369. https://doi.org/10.1016/j.procir.2016.03.144
109. Fitzgerald A et al (2021) A life cycle engineering perspective on biocomposites as a solution for a sustainable recovery. Sustainability 13(3). https://doi.org/10.3390/su13031160
110. Sassanelli C, Rosa P, Rocca R, Terzi S (2019) Circular economy performance assessment methods: a systematic literature review. J Clean Prod 229:440–453. https://doi.org/10.1016/j.jclepro.2019.05.019
111. Hermansson F, Heimersson S, Janssen M, Svanström M (2022) Can carbon fiber composites have a lower environmental impact than fiberglass? Resour Conserv Recycl 181. https://doi.org/10.1016/j.resconrec.2022.106234
112. Hermansson F, Janssen M, Svanström M (2019) Prospective study of lignin-based and recycled carbon fibers in composites through meta-analysis of life cycle assessments. J Clean Prod 223:946–956. https://doi.org/10.1016/j.jclepro.2019.03.022
113. Arnold U, De Palmenaer A, Brück T, Kuse K (2018) Energy-efficient carbon fiber production with concentrated solar power: process design and techno-economic analysis. Ind Eng Chem Res 57(23):7934–7945. https://doi.org/10.1021/acs.iecr.7b04841
114. Das S (2011) Life cycle assessment of carbon fiber-reinforced polymer composites. Int J Life Cycle Assess 16(3):268–282. https://doi.org/10.1007/s11367-011-0264-z
115. Suzuki T, Takahashi J (2005) LCA of lightweight vehicles by using CFRP for mass-produced vehicles. In: Fifteenth international conference on composite materials, Durban, South Africa. Department of Environmental and Ocean Engineering, The University of Tokyo, p 4
116. Griffing E (2009) Carbon fiber from PAN. In: Life cycle inventory summary. Wichita State University
117. Liddell H, Dollinger C, Fischer A, Brueske S, Carpenter A, Cresko J (2017) Bandwidth study on energy use and potential energy saving opportunities in U.S. carbon fiber reinforced polymer manufacturing, USA
118. Liddell H, Brueske S, Carpenter A, Cresko J (2016) Manufacturing energy intensity and opportunity analysis for fiber-reinforced polymer composites and other lightweight materials.

In: Proceedings of the American Society for composites: thirty-first technical conference, Williamsburg, USA
119. Dér A et al (2021) Modelling and analysis of the energy intensity in polyacrylonitrile (PAN) precursor and carbon fibre manufacturing. J Clean Prod 303:127105. https://doi.org/10.1016/j.jclepro.2021.127105
120. Pakdel E, Kashi S, Varley R, Wang X (2021) Recent progress in recycling carbon fibre reinforced composites and dry carbon fibre wastes. Resources, conservation and recycling, vol 166. https://doi.org/10.1016/j.resconrec.2020.105340
121. Das S, Warren J, West D, Schexnayder SM (2016) Global carbon fiber composites supply chain competitiveness analysis. Oak Ridge National Laboratory, The University of Tennessee, Knoxville
122. Duflou JR, Deng Y, Van Acker K, Dewulf W (2012) Do fiber-reinforced polymer composites provide environmentally benign alternatives? A life-cycle-assessment-based study. MRS Bull 37(4):374–382. https://doi.org/10.1557/mrs.2012.33
123. Schmidt JH, Watson J (2013) Eco Island Ferry—comparative LCA of island ferry with carbon fibre composite based and steel based structures, 2.0 LCA consultants, Aalborg, Denmark
124. Suzuki T, Takahashi J (2005) Prediction of energy intensity of carbon fiber reinforced plastics for mass-produced passenger cars. In: Proceedings of 9th Japan International SAMPE Symposium, Japan, 2005: Department of Environmental and Ocean Engineering, The University of Tokyo, pp 14–19
125. Hartmut S (1999) Material intensity of advanced composite materials: Results of study for the Verbundwerkstofflabor Bremen e.V. [Online]. Available: http://hdl.handle.net/10419/49096
126. Gibson TL (2000) Life cycle assessment of advanced materials for automotive applications. SAE Trans 109:1932–1941 [Online]. Available: http://www.jstor.org.ezproxy-b.deakin.edu.au/stable/44687028
127. He D, Soo VK, Kim HC, Doolan M (2021) Life cycle primary energy demand and greenhouse gas emission benefits of vehicle lightweighting with recycled carbon fibre. Procedia CIRP 98:43–48. https://doi.org/10.1016/j.procir.2021.01.003
128. Stergiou V, Konstantopoulos G, Charitidis CA (2022) Carbon fiber reinforced plastics in space: life cycle assessment towards improved sustainability of space vehicles. J Compos Sci 6(5). https://doi.org/10.3390/jcs6050144
129. Ghosh T, Kim HC, De Kleine R, Wallington TJ, Bakshi BR (2021) Life cycle energy and greenhouse gas emissions implications of using carbon fiber reinforced polymers in automotive components: front subframe case study. Sustainable Materials and Technologies, vol 28. https://doi.org/10.1016/j.susmat.2021.e00263
130. Song YS, Youn JR, Gutowski TG (2009) Life cycle energy analysis of fiber-reinforced composites. Compos A Appl Sci Manuf 40(8):1257–1265. https://doi.org/10.1016/j.compositesa.2009.05.020
131. Karuppannan Gopalraj S, Deviatkin I, Horttanainen M, Karki T (2021) Life cycle assessment of a thermal recycling process as an alternative to existing CFRP and GFRP composite wastes management options. Polymers (Basel) 13(24). https://doi.org/10.3390/polym13244430
132. Walker SRJ, Thies PR (2022) A life cycle assessment comparison of materials for a tidal stream turbine blade. Appl Energy 309. https://doi.org/10.1016/j.apenergy.2021.118353
133. Liu S, Han K, Chen L, Zheng Y, Yu M (2015) Influence of air circulation on the structure and properties of melt-spun PAN precursor fibers during thermal oxidation. RSC Adv 5(47):37669–37674. https://doi.org/10.1039/c5ra00476d
134. Wu G, Lu C, Ling L, Hao A, He F (2005) Influence of tension on the oxidative stabilization process of polyacrylonitrile fibers. J Appl Polym Sci 96(4):1029–1034. https://doi.org/10.1002/app.21388
135. Wang B, Li C, Cao W (2021) Effect of stretching on the orientation structure and reaction behavior of PAN fiber during the thermal stabilization. Mater Res Exp 8(8). https://doi.org/10.1088/2053-1591/ac19e9
136. Wang B, Wu S, Li C, Cao W (2021) Evolution mechanism of cyclized structure of PAN-based pre-oxidized fiber during low temperature carbonization process. Mater Res Exp 8(11). https://doi.org/10.1088/2053-1591/ac3587

137. Lee J-E, Choi J, Lee DJ, Lee S, Chae HG (2022) Radial microstructure development of polyacrylonitrile (PAN)-based carbon fibers. Carbon 191:515–524. https://doi.org/10.1016/j.carbon.2022.02.023
138. Khayyam H et al (2015) Stochastic optimization models for energy management in carbonization process of carbon fiber production. Appl Energy 158:643–655. https://doi.org/10.1016/j.apenergy.2015.08.008
139. Terra BM, de Andrade DA, de Mesquita RN (2021) Characterization of polyacrylonitrile thermal stabilization process for carbon fiber production using intelligent algorithms. Polym Test 100. https://doi.org/10.1016/j.polymertesting.2021.107238
140. Paulauskas FL, Sherman DM (2010) Apparatus and method for stabilization or oxidation of polymeric materials. United States Patent 7,649,078 Patent Appl. 11/391,615
141. Paulauskas FL, Naskar AK, Bonds A (2015) Advanced oxidation method for producing high-density oxidized polyacrylonitrile fibers, United States Patent US20120322332A1
142. Naskar AK et al (2017) Magneto-carbonization method for production of carbon fiber, and high performance carbon fibers made thereby, United States Patent Appl. 13/833,834
143. Liu J, Xiao S, Shen Z, Xu L, Zhang L, Peng J (2018) Study on the oxidative stabilization of polyacrylonitrile fibers by microwave heating. Polym Degrad Stab 150:86–91. https://doi.org/10.1016/j.polymdegradstab.2018.02.017
144. Elagib THH, Hassan EAM, Fan C, Han K, Yu M (2019) Microwave pre-oxidation for polyacrylonitrile precursor coated with nano-carbon black. Polym Eng Sci 59(3):457–464. https://doi.org/10.1002/pen.24943
145. Sunil S, Abhilas JK, Kumar A, Shukla HK (2019) Oxidative stabilization studies on pretreated polyacrylonitrile precursor fiber suitable for carbon fiber production. In: Presented at the international conference on inventive material science applications: ICIMA 2019
146. Dang W et al (2020) Structural transformation of polyacrylonitrile (PAN) fibers during rapid thermal pretreatment in nitrogen atmosphere. Polymers (Basel) 12(1):63. https://doi.org/10.3390/polym12010063
147. Lu M et al (2021) Stabilization of polyacrylonitrile fibers with carbon nanotubes. Polym Degrad Stabil 188. https://doi.org/10.1016/j.polymdegradstab.2021.109567
148. Zhao G, Zhang C, Lv L, Liu J, Guo S (2022) Research on dynamic microwave low-temperature carbonization of high performance carbon fiber. Diamond Related Mater 125. https://doi.org/10.1016/j.diamond.2022.108989
149. Geyer R, Kuczenski B, Zink T, Henderson A (2015) Common misconceptions about recycling. J Ind Ecol 20(5):1010–1017. https://doi.org/10.1111/jiec.12355
150. McConnell VP (2010) Launching the carbon fibre recycling industry. Reinf Plast 54(2):33–37. https://doi.org/10.1016/s0034-3617(10)70063-1
151. Patti A, Cicala G, Acierno D (2020) Eco-sustainability of the textile production: waste recovery and current recycling in the composites world. Polymers (Basel) 13(1). https://doi.org/10.3390/polym13010134
152. Zink T, Geyer R (2018) Material recycling and the myth of landfill diversion. J Ind Ecol 23(3):541–548. https://doi.org/10.1111/jiec.12808
153. Geyer R, Jambeck JR, Law KL (2017) Production, use, and fate of all plastics ever made. Sci Adv 3(7):e1700782. https://doi.org/10.1126/sciadv.1700782
154. Sandin G, Peters GM (2018) Environmental impact of textile reuse and recycling—a review. J Clean Prod 184:353–365. https://doi.org/10.1016/j.jclepro.2018.02.266
155. Arafat Y, Uddin AJ (2022) Recycled fibers from pre- and post-consumer textile waste as blend constituents in manufacturing 100% cotton yarns in ring spinning: a sustainable and eco-friendly approach. Heliyon 8(11):e11275. https://doi.org/10.1016/j.heliyon.2022.e11275
156. Hahladakis JN, Iacovidou E (2019) An overview of the challenges and trade-offs in closing the loop of post-consumer plastic waste (PCPW): focus on recycling. J Hazard Mater 380:120887. https://doi.org/10.1016/j.jhazmat.2019.120887
157. Howarth J, Mareddy SSR, Mativenga PT (2014) Energy intensity and environmental analysis of mechanical recycling of carbon fibre composite. J Clean Prod 81:46–50. https://doi.org/10.1016/j.jclepro.2014.06.023

158. Guo W, Bai S, Ye Y, Zhu LA (2019) Recycling carbon fiber-reinforced polymers by pyrolysis and reused to prepare short-cut fiber C/C composite. J Reinforc Plast Compos 38(7):340–348. https://doi.org/10.1177/0731684418822144
159. Okajima I, Sako T (2015) Recycling of carbon fiber-reinforced plastic using supercritical and subcritical fluids. J Mater Cycles Waste Manage 19(1):15–20. https://doi.org/10.1007/s10163-015-0412-9
160. Yu K, Shi Q, Dunn ML, Wang T, Qi HJ (2016) Carbon fiber reinforced thermoset composite with near 100% recyclability. Adv Func Mater 26(33):6098–6106. https://doi.org/10.1002/adfm.201602056
161. Arif ZU, Khalid MY, Ahmed W, Arshad H, Ullah S (2022) Recycling of the glass/carbon fibre reinforced polymer composites: a step towards the circular economy. Polym-Plast Technol Mater 61(7):761–788. https://doi.org/10.1080/25740881.2021.2015781
162. Qazi H, Subiantoro A, Lin R, Jayaraman K (2022) Prediction model of optimised process parameters for recycling carbon fibre composites. Therm Sci Eng Progr 33. https://doi.org/10.1016/j.tsep.2022.101344
163. Giorgini L et al (2015) Recovery of carbon fibers from cured and uncured carbon fiber reinforced composites wastes and their use as feedstock for a new composite production. Polym Compos 36(6):1084–1095. https://doi.org/10.1002/pc.23440
164. Kratz J, Low YS, Fox B (2017) Resource-friendly carbon fiber composites: combining production waste with virgin feedstock. Adv Manuf Polym Compos Sci 3(4):121–129. https://doi.org/10.1080/20550340.2017.1379257
165. Harper International (2020) Direct fired thermal oxidizer (DFTO). Harper International. Accessed 16 Mar 2020
166. Harper International (2020) Regenerative Thermal Oxidizer (RTO). Harper International. Accessed 16 Mar 2020
167. Harper International (2020) Waste gas treatment & handling. Harper International. Accessed 16 Mar 2020
168. A. E. Systems (2020) Carbon fibre: furnace & oven emission control. Anguil Environmental Systems. Accessed 16 Mar 2021

Chapter 3
Materials and Methodology

This work is divided into two parts, the first part aims to develop an LCA and LCC-based assessment methodology, tailored to carbon fibre production, which is flexible enough to adapt to various production circumstances, such as various precursor materials, but at the same time easy to use and yields comparable results. The second part employs the developed assessment methodology to analyse the carbon fibre manufacturing process in-depth and effects of changes of process and material parameters. This chapter first gives a short overview of the methodological foundation of the assessment concept, complementing information already detailed in the review section of this work. Subsequently used material and the manufacturing equipment including the data acquisition system are introduced. Finally, the material characterisation techniques, utilized in this study, are detailed.

The large scale of the experimental second part of this study only permitted a very limited number of experiments for each investigation. Overall, the work was therefore designed in such a way as to keep manufacturing runs as short as possible while yielding as much significantly distinct information as possible. To define initial manufacturing parameter sets for optimization, expert knowledge from the Carbon Nexus operations team in combination with, albeit scarce, literary references, were employed. The manufacturing process was subsequently carefully optimized to meet set requirements, such as carbon fibre material properties, material throughput or temperature profiles.

3.1 LCA and LCC Methodology

Raising awareness of the environmental impact of products has led to the development and increased application of a broad range of holistic assessment methodologies over the last decades. Life cycle assessment (LCA) is an important tool to evaluate a product's impact and interaction with its environment throughout its life cycle. The

T. Groetsch, *Enhancing Carbon Fibre Production Through Life Cycle Analysis*, Springer Theses, https://doi.org/10.1007/978-3-031-70578-6_3

basic idea is to identify, quantify and assess environmental impact from the very start (cradle), including raw material extraction, to the very end (grave), i.e. disposal/recycling and everything in between (gates), such as production, use and end of life treatment. Studies can be conducted cradle to grave, cradle to gate, gate to gate or gate to grave. The methodology is internationally standardized in ISO 14040:X [1] and 14044:X [2] with the former more generally outlining the methodological approach and the wider framework and the latter focusing on more practical requirements and guidelines. Overall, these standards only outline basic concepts, principle and approaches, which leaves room for interpretation, adjustment to specific use-cases and improvement of the methodology itself.

LCA's typically consist of the four phases shown in Fig. 2.8 in the review section. The four standard phases are used as a basis for the development of the streamlined concept discussed in Chap. 4. Standardized goal and scope, product unit and reference flow are used to ensure comparability. As carbon fibre processing as a number of peripheral processes, the system boundaries are narrowly set to only include key process steps. For the second phase inventory analysis, standard datasets to be collected are defined with all key contributors represented by measured values not database quantities. In the subsequent impact assessment phase the collected data is translated into major environmental impacts, classified and summarized into key impact categories based on the process's location in Waurn Ponds, Australia. The assessment finally produces a standardized report for interpretation of the generated results. To address a key issue of today's carbon fibre, cost, based on LCC principles, are incorporated in the assessment. The concept and the approaches used are outlined in detail in Chap. 4.

3.2 Reference Material

For the most part of this work, material related experiments were performed using a commercial 24,000 filament (24k) PAN precursor. The chemical composition of the precursor includes acrylonitrile (AN), methyl acrylate (MA), itaconic acid (IA) with a silicon oil-based spin finish applied for improvement of handling. Only for Chap. 7 a different reference material was used in that three tows of 12,000 filaments (12k) were processed. The chemical composition however remained the same. For all studies included in this work, the precursor material was processed under conditions akin to commercial production, when scaled to size. For the resultant carbon fibres, material parameters commonly referenced as commercially relevant grade of carbon fibres were always targeted [3, 4]. These properties were chosen based on typical properties of commercially available high quality, low-cost carbon fibres mainly intended for automotive or wind turbine applications [5, 6]. The targeted values for the main material properties were a tensile strength of > 3.0 GPa, an initial tensile modulus of > 200 GPa and mass density in the range of 1.7–1.9 g/cm^3. In Chap. 4 two other materials, a textile-grade PAN precursor and a Cellulose-Lignin blend are analysed, details of which are outlined in the specific chapter.

3.3 Manufacturing Equipment

For this study, most of the data was collected from the research line at Carbon Nexus, Deakin University (Australia), a scaled down version of a typical commercial production setup. The production line consists of four stabilization ovens with a maximum of three passes per oven. This enables stabilization times in the range of typical processing times of $\approx$ 50–80 min as found in industrial production. The carbonization furnaces consist of three (LT) and two (HT) separate zones with individually controlled increasing temperatures. Both furnaces operate with an inert atmosphere employing a constant flow of nitrogen. The electrolytic surface treatment and sizing section consists of a series of baths with the required liquids for each step constantly circulating. The material subsequently passes through non-contact dryers before being wound onto paper-cores. The whole process is designed as a continuous conversion of the material, with speed and tension controlled by a series of roller-drive systems between the individual process steps. Emissions generated in the furnaces are removed via two incinerators breaking down the most hazardous substances. All gases are removed through a fan driven exhaust stack. Figure 3.1 shows a schematic representation of the continuous carbon fibre manufacturing line utilized for this work with the process steps outlined above.

The detailed specifications of the carbon fibre research line utilized throughout this work are shown in Table 3.1 [8, 9]. Of note here is, that the equipment slightly varies in size, i.e. the stabilization ovens have a higher theoretical maximum capacity than the furnaces. Especially energy-related analyses comparing data within the process are therefore more difficult and need to take the different sizes into account for better accuracy. As this work mainly compares different overall processing conditions, percentual changes between these trials however can well be used for comparison.

There is a second, significantly larger, production line available at Carbon Nexus. The basic principles in terms of processing are the same as for the line described above. The larger pilot line can however process up to 30 tows of 24,000 filaments simultaneously and is equipped with a direct fired thermal oxidizer for abatement of emission. Due to the large scale and the high cost of operation, data from this line was only used for the initial verification of the developed assessment methodology and to showcase that the concept is easily adaptable to different manufacturing scales.

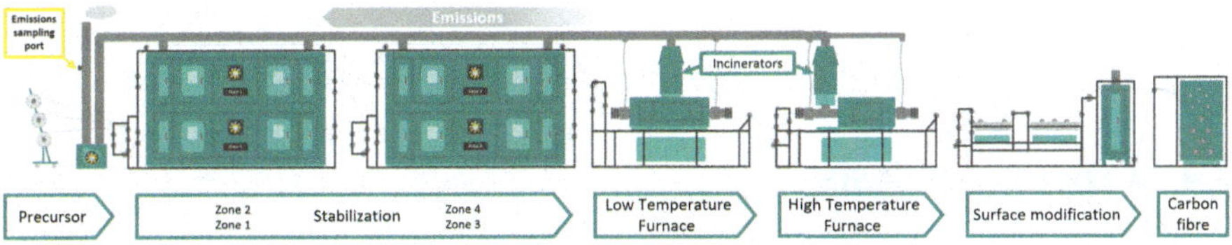

Fig. 3.1 Schematic representation of the carbon fibre research line at Carbon Nexus. Reproduced from [7] with permission from Elsevier © Elsevier 2023

Table 3.1 Carbon fibre research line specifications and manufacturing details

Process stage	Parameters (units)	Value
General		
	Reference line speed range (m/h)	15–60
	Typical number of tows	Up to 6 (24 k)
Stabilization		
	Number of stabilization zones/ovens	4
	Number pf passes per zone/oven	3
	Maximum operating temperature (°C)	400
	Total heated length (m)	24
	Heating power per zone (KW)	40
	Recirculation fan power (1 per Zone) (KW)	5.6
	Air flow delivery system	Centre-to-end
Low temperature carbonization (LT)		
	Number of carbonization zones	3
	Maximum operating temperature (°C)	1100
	Total heated length (mm)	1500
	Heating power (KW)	33
	Typical nitrogen supply rate (m^3/h)	15
	Cooling system technology	water
High temperature carbonization (HT)		
	Number of carbonization zones	2
	Maximum operating temperature (°C)	1600
	Total heated length (mm)	1000
	Heating power (KW)	40
	Typical nitrogen supply rate (m^3/h)	15
	Cooling system technology	water
Finishing treatments		
	Surface treatment—electrolytic cell	Max. 200 A @ 30 VDC (variable)
	Surface treatment -electrolyte	Ammonia Bicarbonate
	Sizing	Michelmann Hydrosize® EP83
	Dryers	Non- contact, radiant (max. 180 °C)

3.4 Process Analysis

The production process is continuously monitored by a host of sensor and data provided by the manufacturing equipment. Additionally manual measurements are taken and material samples are tested throughout trials to ensure consistent processing. Key material parameters, processing conditions and measurements are additionally recorded on a process specification sheet for reference.

3.4.1 Process Parameters

The production line is equipped with a supervisory control and data acquisition (SCADA) system to control and monitor the manufacturing process. The system gathers and stores real-time data for more than 120 process parameters. Most critical these include set, warning and limit temperatures and drive speeds besides power related information and device specific state data. Process parameters can be changed via a touch-pane displaying all relevant live information and related messages and warnings. The programmable logic controller continuously records all parameters in 12h sequences, even when the line is idle, and stores them on the internal hard drive. For safety and data security there is no direct network access to the data, although technically possible. Relevant data files have hence to be manually transferred to a network location for further analysis. As only live currents are recorded for several devices, the energy demand must be calculated afterwards. Most notably no currents or energy related data is recorded for the heating elements in the ovens or furnaces, which is why additional current transducers were installed and data recorded concurrently.

The control system of the research line records the real time power demand/current for various consumers such as circulation fans or drive motors. Depending on the equipment's electric wiring, Eq. 3.1 (three phase) and 2 (single phase) were used to calculate the electric power:

$$P = \frac{\sqrt{3} \cdot PF \cdot I \cdot U}{1000} \text{ [kW]} \tag{3.1}$$

where P is the power (kW), PF is the power factor, here assumed to be 1, I is the current (A) and U the voltage (V).

$$P = \frac{I \cdot U}{1000} \text{ [kW]} \tag{3.2}$$

where P is the power (kW), I is the current (A) and U is the voltage (V).

During stable state production, recorded data typically showed no significant fluctuation in recorded power or current, hence the energy demand was simply calculated using Eq. 3.3.

$$E = P \cdot h \text{ [kWh]} \tag{3.3}$$

where E is the energy demand in kWh, P is the power (kW) and h is the time (hrs).

For this work the focus lay on energy related data and to a lesser extent on temperatures. Other parameters were not directly utilized and only monitored for fluctuations or deviations throughout the trials.

3.4.1.1 Current Transducers

As the SCADA system does not capture data from the heating elements of the ovens and furnaces, additional measurement devices were installed to evaluate the equipment. Fastron AT 150 B10 self-powered split core AC current sensors with average RMS output (0–10 V) were attached to the connections supplying electricity to the heating elements to record drawn current and subsequently calculate the energy demand of the equipment. The data was recorded using a Hioki Memory HiCorder MR8875 at a rate of 1Hz and energy consumption was subsequently calculated using the formula shown above.

3.4.2 Gas Fourier Transform Infrared Spectroscopy (FTIR)

The conversion process of PAN to carbon fibres emits a number of hazardous gases, among other hydrogen cyanide, ammonia and methane [10, 11]. The chemical composition of the evolving gases was determined using a portable Gasmet DX 4000 FT-IR gas analyser (Gasmet Technologies Oy, Finland) equipped with a temperature-controlled sampling unit (APP100). Using the associated Calcmet software (Gasmet Technologies Oy, Finland), individual gases were identified and the gas concentrations were quantified by comparing the generated infrared absorption spectra against reference libraries of gases of known concentrations [12–14]. A modified classical least squares (CLS) fitting algorithm combined with the assumption that absorbance is proportional to concentration within calibration limits [15] was used to quantify the composition of gaseous emissions [16]. An overall reference spectrum is calculated from the sum of relevant single gas reference spectra weighted by individual adjustment coefficients. Subtraction of the two overall spectra (reference and sample) generated a residual spectrum, indicating the accuracy of the curve fitting procedure performed by the software. Ideally, the residual spectrum had no significant patterns or shapes in the analysed areas with only low levels of remaining noise. To reduce overlap/interference between compounds, a spectral range is defined for each analyte within which the algorithm compares and fits the sample spectrum with the relevant reference spectrum. To reduce overall interference, water vapour and CO_2 are included in the reference library and treated like all other analysed substances [16].

Prior to taking measurements, the sampling unit system was pre-purged with nitrogen and heated to a temperature of 180 °C. A background spectrum of the production plants standard atmosphere was recorded as a baseline prior to each trial. To take the measurements, the sampling probe was inserted into the middle of the exhaust stack duct (ø 150mm) using the sampling port shown in Fig. 3.1. Each gas sampling experiment was conducted at least 2 times. The gases were sampled at least 15 times for a duration of one minute for each experiment. Before starting to record measurements, at least 5 samples were compared for stability of values.

The gas flow velocity and temperature of the gases were measured with an anemometer (Kanomax Anemomaster 6162). Measurements were performed for a total duration of 5 min before and after each gas emission measurement. These measurements were necessary to relate the sampled emissions to specific material quantities produced.

3.4.2.1 Mass Flow Calculations

The concentration of gases calculated by the Gasmet DX4000 FT-IR gas analyser is provided in parts per million (ppm). To convert the concentration, $c_{(\mathrm{ppm})}$, of each gas to g/m^3, the following Eq. 3.4 was used:

$$c = \frac{c_{(ppm)} \cdot M \cdot 10^{-3}}{V_T} \left[\frac{\mathrm{g}}{\mathrm{m}^3}\right] \tag{3.4}$$

where c is the concentration of the individual gas in g/m^3, M is the molar mass of the gas expressed in g/mol and V_T is the volume of one mole of gas at a specific temperature in litre.

To calculate the volume V_T of one mole of the gas at temperature, the ideal gas law, shown in Eq. 3.5, was applied [17]:

$$V_T = \frac{n \cdot R \cdot T \cdot Z}{P} \left[\mathrm{m}^3\right] \tag{3.5}$$

where n is the number of moles of gas, here equal to 1, R is the Boltzmann constant, which is 8.2057338×10^{-2} L atm mol^{-1} K^{-1}, T is the temperature in Kelvin, P is the pressure of the gas expressed in atmosphere. Z is the compression factor of the gas mixture, here set to 1, as for calculation purposes the gas mixture was considered to be ideal in line with remarks in ISO 14912-2006.

The volumetric flow rate ($\dot{\mathrm{V}}$) was calculated using Eq. 3.6, where v is the flow velocity and A the cross-sectional area of the pipe in m^2. Here the flow velocity is measured with a flow meter and expressed in meters per second.

$$\dot{V} = A \cdot v \left[\frac{\mathrm{m}^3}{\mathrm{s}}\right] \tag{3.6}$$

The mass flow rate ($\dot{\mathrm{m}}$), g per hour, was then calculated using Eq. 3.7:

$$\dot{m} = \dot{V} \cdot c \cdot 3.6 \left[\frac{\mathrm{g}}{\mathrm{h}}\right] \tag{3.7}$$

The furnaces of the Carbon Nexus research line are equipped with incinerators to break down emitted gases. For the purpose of this work, emissions were measured unincinerated. As this type of emission treatment is not used on an industrial scale, where thermal oxidizers with or without heat recovery are common, this was thought to be the best way of conveying the desired information about changes in emitted gases.

3.4.3 *Various Manual Measurements*

A few process parameters, though important, are not digitally recorded, but manually measured and constantly checked while the line is operating. The tension of the material is not recorded by the SCADA system but is critical to control output material properties. Analog and digital hand-held tension-meters from Hans Schmidt & Co, Germany are used for this purpose and values are recorded in the process specification sheets.

The air intake of ovens is controlled through the width of the slots where the fibres enter and exit. The setting is checked and adjusted with the help of smoke pens based on experience.

Similarly the furnace atmosphere or more specifically the slight overpressure of the inert atmosphere is checked at the end of the furnaces, ensuring that no air enters the champers. Nitrogen flow can be adjusted via a simple analogue inline flowmeter.

3.5 Material Characterisation

The source material, intermediate products and the final carbon fibres were characterised using a number of techniques. Of the methods described below, the most important ones used for every part of this work, were the following three: single fibre tensile testing, mass density measurements and FTIR spectroscopy to calculate the extent of cyclisation reaction. All three techniques were used to optimize each trial as well as for the final comparison of material produced.

3.5.1 *Single Fibre Tensile Test*

To determine the mechanical properties of the used and produced fibre material, single fibres were subjected to tensile tests. The single filament material properties of the precursor fibres, intermediate fibres and final carbon fibres were measured using a Favimat + single fibre tensile tester (Textechno, Germany) with a 210 cN load cell and clamps with a surface area of $4 \times 4\ mm^2$. The tensile strength and initial modulus of a minimum of 50 individual filaments were measured at a pretension of 0.5N. The tests were conducted at 8 mm/min and a gauge length of 20 mm for precursor and stabilized fibres and 2 mm/min and a gauge length of 25mm for carbon fibres. The instrument automatically calculates the linear density using a vibroscopic method. This method measures the resonance frequency of the sample at known gauge length and pre-tension and subsequently calculates the linear density in accordance with ASTM D 1577: Standard Test Methods for Linear Density of Textile Fibers - Option C: Vibroscope, General.

3.5.2 *Mass Density*

The mass density of PAN precursor fibres, stabilized fibres and carbon fibres were measured at a standard 23 °C using a density gradient column in accordance with ASTM D1505-10: Standard Test Method for Density of Plastics by the Density-Gradient Technique. One column was filled with a calcium nitrate and distilled water solution to characterize precursors and stabilized fibres with a density gradient from 1.17 to 1.42 g/cm^3. Another column was filled with a solution of 1,3-dibromo-propane and triethyl-phosphate to characterize carbon fibres with a density gradient from 1.6 to 1.9 g/cm^3.

3.5.3 *Linear Density*

The weight of the fibre was measured after each manufacturing stage. Fibre tow samples of 1m length were carefully cut to measure the linear density, where the weight of the fibre samples was measured 3 times each. These measurements were then used to obtain the overall yield of the process.

3.5.4 *Fourier Transform Infrared Spectroscopy (FTIR)*

The evolution of the fibre chemistry during the stabilization stage was measured using a Bruker Lumos FTIR spectrometer equipped with a germanium crystal. Each sample was measured using the attenuated total reflectance (ATR) mode at four spots along the tow to verify the homogeneity of the sample at room temperature without prior drying. Every measurement was taken from 600 to 4000 cm^{-1} with an average of 64 scans at a resolution of 4 cm^{-1}. The evolution of stabilization was quantified using the extent of cyclization reaction (EOR) index (Eq. 3.8) [18]:

$$EOR(\%) = \frac{\left(100 \cdot 0.29 \cdot Abs_{(1590)}\right)}{\left(Abs_{(2242)} + \left(0.29 \cdot Abs_{(1590)}\right)\right)} \tag{3.8}$$

where $Abs_{(1590)}$ and $Abs_{(2242)}$ were the measured absorbance of the C=C, C=N and C≡N functional groups, respectively.

3.5.5 *Elemental Analysis*

CHNS elemental analysis of fibre samples was performed using a Vario Macro Cube Elemental analyser (Elementar, Germany). All samples were tested at least three times. The carbon (C), hydrogen (H), and nitrogen (N) content were determined directly, whereas the oxygen (O) content was calculated by difference. Prior to the elemental analysis measurement, the potential contaminants and spin finish (silicon oil based) were removed by carefully washing the fibres in acetone solutions.

3.5.6 *Thermogravimetric Analysis (TGA)*

The thermal degradation of precursor material was determined using a thermogravimetric analyser TG 209 F1 Libra (Netzsch, Germany). Approximately 5mg of fibre material were weighed and placed in an aluminium oxide crucible. The sample was subsequently placed in the furnace chamber of the instrument, which was either purged with 20 ml/min Air or 20 ml/min Nitrogen, depending on the temperature range the analysis was performed. The flow gas was chosen based on the atmosphere the fibre material would be processed in at specific temperatures. The gas used was therefore switched at the temperature the fibre would go from stabilization to carbonization (typically around 260 °C). Transition temperatures were subsequently determined using the accompanying Proteus software (Netzsch, Germany).

References

1. ISO 14040 (2006) Environmental management—life cycle assessment—principles and framework, I. O. f. Standardization, Geneva, Switzerland [Online]. Available: https://books.google.com.au/books?id=C_xSygEACAAJ
2. ISO 14044 (2006) Environmental management—life cycle assessment—requirements and guidelines, I. O. f. Standardization, Geneva [Online]. Available: https://books.google.com.au/books?id=1SEkygAACAAJ
3. Park S-J, Heo G-Y (2015) Precursors and manufacturing of carbon fibers. In: Park S-J (ed) Carbon fibers. Springer Series in Materials Science. Springer Netherlands, Dordrecht, Chapter 2, pp 31–66
4. Inagaki M (2000) New carbons—control of structure and functions. Elsevier, p 229
5. Hiremath N, Young S, Ghossein H, Penumadu D, Vaidya U, Theodore M (2020) Low cost textile-grade carbon-fiber epoxy composites for automotive and wind energy applications. Compos Part B: Eng 198. https://doi.org/10.1016/j.compositesb.2020.108156

6. Ennis BL et al (2019) Optimized carbon fiber composites in wind turbine blade design. Sandia National Lab (SNL-NM), Albuquerque, NM, USA
7. Groetsch T, Maghe M, Creighton C, Varley RJ (2023) Environmental, property and cost impact analysis of carbon fibre at increasing rates of production. J Clean Prod 382. https://doi.org/10.1016/j.jclepro.2022.135292
8. Maghe M et al (2016) Using ionic liquids to reduce energy consumption for carbon fibre production. J Mater Chem A 4(42):16619–16626. https://doi.org/10.1039/c6ta06842a
9. Nunna S et al (2019) Time dependent structure and property evolution in fibres during continuous carbon fibre manufacturing. Materials (Basel) 12(7):1069. https://doi.org/10.3390/ma12071069.
10. Morgan P (2005) Carbon fibers and their composites. CRC Press, Ch. 1174, pp 185–259
11. Chung DDL (1994) Carbon fiber composites (no. 1). Heinemann, Newton, p 209
12. Bacsik Z, Mink J, Keresztury G (2004) FTIR spectroscopy of the atmosphere. I. Principles and methods. Appl Spectrosc Rev 39(3):295–363. https://doi.org/10.1081/asr-200030192
13. Fu P, Yi W, Bai X, Li Z, Hu S, Xiang J (2011) Effect of temperature on gas composition and char structural features of pyrolyzed agricultural residues. Biores Technol 102(17):8211–8219. https://doi.org/10.1016/j.biortech.2011.05.083
14. Bak J, Clausen S (2002) FTIR emission spectroscopy methods and procedures for real time quantitative gas analysis in industrial environments. Meas Sci Technol 13(2):150–156. https://doi.org/10.1088/0957-0233/13/2/302
15. Laakso O et al (2000) The use of low-resolution FT-IR spectrometry for the analysis of alcohols in breath. J Anal Toxicol 24(4):250–256. https://doi.org/10.1093/jat/24.4.250
16. Ahonen I, Riipinen H, Roos A (1996) Portable Fourier transform infrared spectrometer for use as a gas analyser in industrial hygiene. Analyst 121(9):1253–1255. https://doi.org/10.1039/an9962101253
17. ISO 14912:2003 (2003) Gas analysis—conversion of gas mixture composition data, I. O. f. Standardization, Geneva, Switzerland
18. Collins GL, Thomas NW, Williams GE (1988) Kinetic relationships between heat generation and nitrile consumption in the reaction of poly(acrylonitrile) in air at 265°C. Carbon 26(5):671–679. https://doi.org/10.1016/0008-6223(88)90070-x

Chapter 4
Development of a Streamlined LCA/LCC Modelling Concept for Carbon Fibre Manufacture

For the widespread application of carbon fibre in CFRP's, there is one key hurdle besides the need for a proper circular recycling strategy. The production of carbon fibres consumes substantially more energy than a comparable amount of steel or aluminium, leading to higher cost and related environmental impacts [1, 2]. Furthermore, the chemical transformation occurring during the production process produces significant amounts of hazardous and environmentally harmful emissions. As the mass of the fibres is reduced by approximately 50% from precursor to carbon fibre during the process depending on material and process parameters, gases such as carbon monoxide, carbon dioxide, hydrogen cyanide and ammonia are released [4]. Through waste gas treatment the most toxic substances can be reduced, but overall emission of greenhouse gases, continue to rise [4, 5].

4.1 Requirements

As outlined in previous chapters, LCA is a well-established and standardized analysis methodology to evaluate the environmental impacts of product systems [6, 7]. Advantages of LCA are its systemic life cycle perspective and using laws of science to provide quantitative information that can be integrated into engineering tools like simulations or models [8]. Detailed assessments, however, typically require a high level of expert knowledge both of the methodology and the products or processes to be analysed. Furthermore full-scale cradle-to-grave analyses can be very time-consuming and therefore expensive and difficult to perform. Life cycle engineering (LCE) and life cycle sustainability assessment (LCSA) utilize LCA and other tools to enable a global view on product systems and their development from an engineering perspective incorporating environmental, economic and social angles [9–11]. This

---

This chapter (modified) was published in [3], Copyright Elsevier 2021.

T. Groetsch, *Enhancing Carbon Fibre Production Through Life Cycle Analysis*, Springer Theses, https://doi.org/10.1007/978-3-031-70578-6_4

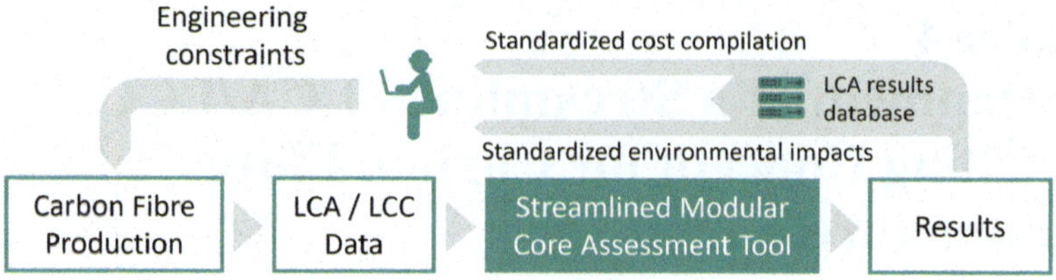

Fig. 4.1 Assessment approach based on [9–11]. Reproduced from [3] with permission from Elsevier © Elsevier 2021

holistic approach fosters the application of LCA in line with engineering activities throughout a products lifecycle. Fig. 4.1 illustrates a potential application scenario for a tailored evaluation approach for carbon fibre production in the context of LCE. A streamlined assessment tool is used in an iterative way to quickly generate standardized results, based thereon (re-)define constraints for the production process and subsequently generate new data for the assessment tool. The aim of an approach like this is to reduce overall complexity and accelerate LCA application, while adhering to accepted principles [12, 13]. A more straightforward standartized concept lowers the barrier of entry for process engineers and material researchers to utilize and apply LCA methodology without the need for extensive training or prior knowledge. As such a concept is firmly based on internatinally accepted standards, comparability with other assesmemts, general acceptance by specialists can be expected to be high.

Therefore, a streamlined assessment is proposed to speed up studies and reduce the extent of required expert knowledge. Besides the environmental impact, high material cost is one of the main inhibitors for a more widespread adoption of carbon fibres [14]. Hence, cost and environmental impact should be analysed in an integrated approach. Incorporating a cost perspective, the approach offers a more holistic view of the production process and addresses the key issue of today's carbon fibre production. The concept is mainly intended to be used in early development phases by material researchers and process developers with limited resources and experience in conducting LCA studies. In addition to tailoring LCA to the carbon fibre production process, an assessment database is proposed to help compare results and steer research towards environmental and economically viable improvements.

4.2 Background

Carbon fibre production is a continuous process consisting of two major stages – first the production of precursor materials and second the conversion of precursor materials to carbon fibres. The chapter focuses on the second stage, as illustrated in Fig. 4.2. The conversion procedure, both from a chemical and process perspective is describe in the review section of this manuscript. There are several additional peripheral processes, which do not directly add value. As Dér et al. [15] showed,

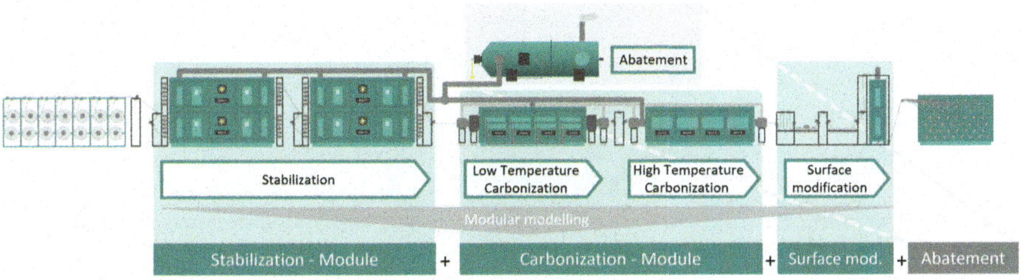

Fig. 4.2 Overview of conversion process of precursor material to carbon fibre and related assessment modules. Reproduced from [3] with permission from Elsevier © Elsevier 2021

most of these processes require only low amounts of electric energy compared to the equipment directly involved in converting the fibres or are static contributors independent of minor process changes, such as building services. As an exception the abatement system, which significantly influences the energy demand, should be considered when determining environmental impacts [15].

4.2.1 Energy Demand of Carbon Fibre Production

While a number of LCA-based studies of carbon fibre production have been published, evaluation and comparison prove difficult. Goals and scopes vary and information on production setups are often not given in detail. Section 2.5 and Table 2.1 detail various reported assessment results and are compared based on the stated energy demand per kg carbon fibre, depicting the scale of variance found in published studies.

4.2.2 Combined LCA and LCC

LCA combines energy balances with other product flows like raw material and waste flows to determine a product's environmental impact. Besides the environmental impact, high cost is one of the main inhibitors for a more widespread adoption of carbon fibres [14]. Therefore, these two factors should be analysed using an integrated approach as outlined in the introductory part of this chapter. As outlined in Sect. 2.4.2 LCC was developed to assess costs incurred by or associated with a product throughout its life cycle [16]. Numerous concepts have been developed which combine LCA and one of the LCC variants into interdisciplinary approaches. As the proposed concept is meant to be a very streamlined approach aiming for a straightforward application without intensive prior training, the scope of the incorporated LCC methodology is limited, focusing on internal costs, specifically of material and energy. It is therefore only loosely based on reviewed approaches [17–19]. Fig. 4.3

Fig. 4.3 Streamlined LCA and LCC structure based on [7, 14, 21]. Reproduced from [3] with permission from Elsevier © Elsevier 2021

shows the proposed four step approach based on LCA principles with joint phases one and two. From a methodical perspective, impact assessment and interpretation are conducted separately, as the scope of the assessments slightly differ. While the LCA side can exclude upstream processes, as outlined in the goal and scope definition of the assessment, they are monetarized in LCC through the input material cost and energy prices. The combination of the two methodologies can help to analyse both environmental and financial impacts with reduced effort. To efficiently combine the two approaches, literature discusses a number of streamlining methods for LCA and LCC [20].

4.3 A Modular LCA/LCC Modelling Concept

Carbon fibre production covers a number of core steps with usually no forks or by-products. It is therefore reasonable to base the proposed modular assessment approach on representations of these core steps. Based thereon, procedures and challenges of streamlining the methods for carbon fibre production are discussed.

4.3.1 Modelling Architecture

The proposed concept in summary is based on the principles of LCA, incorporating the basic evaluation of costs through LCC techniques. The combination of simplification and interdisciplinarity is tailored towards fast and easy comparison of various scenarios in a research environment. It is intended to support the development of new materials and processes as well as optimization thereof, from an early stage to streamline the research process considering environmental /economic feasibility. The modularity (Fig. 4.2) of the concept ensures comparability that is straightforward to verify and simplifies adaption of the model to specific production setups.

On its own, the methodology can only provide parts of the overall assessment results, but through its standardized form, these outcomes are comprehensibly comparable. The basic concept of the proposed core model is shown in Fig. 4.4. Based on a standardized scope and system boundaries, life cycle inventory data is gathered and subsequently aggregated in three impact assessment modules, energy, environmental impact and cost.

Figure 4.5 shows the environment in which the modular core assessment is intended to be embedded. As a core concept, it is not meant to replace existing methodology, but rather be integrated in assessment frameworks and used case by case. Consequently, assessments can adhere to standards, be tailored to specification and at the same time produce comparable results. As the system boundaries of the core assessment are defined by specific production steps/equipment and are divided into standardized modules, the overall scope of the assessment can easily be extended to include up- or downstream production processes, adhering to the defined principles.

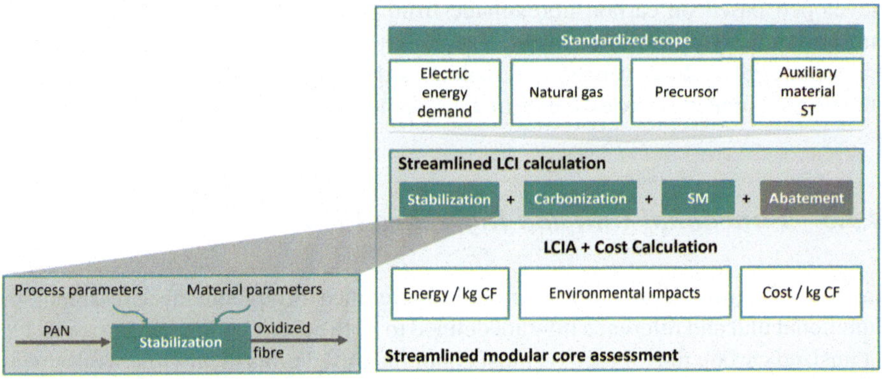

Fig. 4.4 Proposed streamlined core model based on LCA and LCC principles. Reproduced from [3] with permission from Elsevier © Elsevier 2021

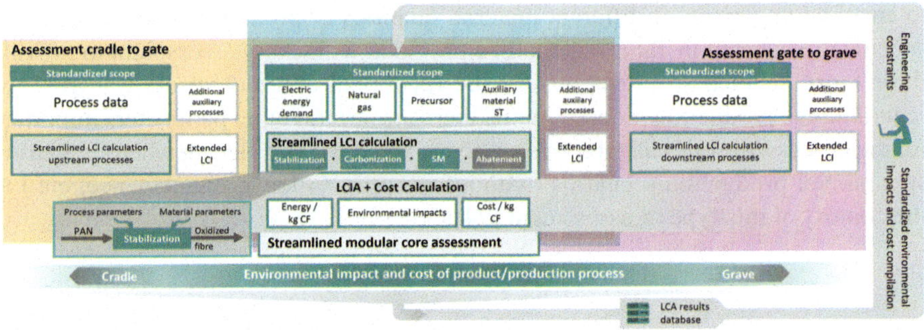

Fig. 4.5 Broader methodical concept of the proposed modelling approach. Reproduced from [3] with permission from Elsevier © Elsevier 2021

4.3.2 Goal and Scope

The approach is intended to assist material researchers and process engineers with evaluating changes in environmental impact and cost. The changes are based on measures derived from research and engineering activities, such as changes of process parameters, innovations in the conversion process or new materials. As the concept aims at streamlining the comparison of these variations, the influence of ramp-up, ramp-down and maintenance times are ignored and only stable state production is evaluated. To tailor the assessment to carbon fibre production and streamline it, the product system, its functions, the functional unit, system boundaries and impact categories ("climate change", "acidification" and "eutrophication") are specified by the model itself and do not have to be defined for each assessment individually. The concept groups repetitive production steps (e.g. stabilization zones or ovens) into modules (see Fig. 4.2) to simplify the overall setup. This standardisation is outlined in the goal and scope definition. The same logic can be applied when extending the assessment to include up- or downstream production processes. This concept was developed based on carbon fibres made from polyacrylonitrile (PAN), as it is the most commonly used source material in commercial production today. Adaption to other materials is however easily possible, as the overall process is very similar for most source materials available or in development today.

4.3.3 Functional Unit and Reference Flow

In order to compare results and to provide a reference throughout the assessment, the functional unit and reference flow are defined to be the conversion of precursor to 1 kg of unsized carbon fibre, based on most commonly used units in previous assessments [1, 4, 6, 7, 22, 23]. Typically previous assessments specified the technical properties in more detail, in line with what LCA methodology suggests [6, 7]. However as this concept aims at comparing production scenarios in early research phases, the functional unit is not restricted by detailed specifications in terms of material properties to enable comparison of different material qualities or production parameters. The specifications listed in Table 4.1 shall be provided with every assessment to provide these more detailed specifications for the functional unit and ensure transparency regarding the production process. While there are several other parameters important for the process, the ones listed above are seen to be the most relevant when comparing streamlined production scenarios (without ramp-up / -down, maintenance, etc.) in the context of this type of assessment [24].

Table 4.1 To be specified parameters of assessed product/production

Parameters	Values (exemplary)
Material	
Precursor type	e.g. PAN, Lignin...
Precursor reference	Manufacturer, Name
Number of tows and size	e.g. 30 × 50 k
Modulus carbon fibre	e.g. 350 GPa
Strength carbon fibre	e.g. 3 GPa
Process	
Production line working width	e.g. 5000 mm
Reference line speed	e.g. 200 m/h
Overall yield	e.g. 0.45
Residence times per step	e.g. Oxid.: 90 min, Carb.: 10 min
Yield per step	e.g. Oxid.: 0.95, Carb.: 0.60
Heated length per step	e.g. Oxid.: 500 m, Carb.: 10 m
Abatement type (per step)	e.g. Direct Fired Thermal Oxidizer

4.3.4 System Boundaries and Cut-Off Criteria

Elements below a threshold percentage or a specific cut-off amount may be omitted if their overall contribution to the assessment results is limited [6, 7]. Because of the rather straightforward production setup, the scope and system boundaries can be standardized for all assessments of this type. They need to be set in a way that is easily reproducible, while also be flexible enough to adjust to different scenarios and include relevant equipment and processes. Therefore, only three basic production steps or modules (stabilization, carbonization and surface modification), as outlined in Fig. 4.2, are evaluated. Additionally, the waste gas abatement system, due to its high energy demand and direct influence on the environmental impact of the production, must be part of the assessment. Any other peripheral processes shall be excluded.

Omitting up- and downstream processes can greatly reduce the necessary amount of data and only has a very limited impact on the quality of the studies, especially for comparative assessments. Nevertheless, the potential influence of process parts to be disregarded has to be thoroughly analysed and clearly stated in the assessment [25]. As the proposed core assessment is a gate-to-gate evaluation of the carbonization process, it only includes the manufacturing steps mentioned above. Up- and downstream processes are not taken into consideration from an LCA perspective. Costs for input flows are however considered in the economic part of the assessment.

4.3.5 Life Cycle Inventory

For the assessed production processes, data like energy consumption, material throughput and specific settings should be directly collected. Relevant production equipment and processes are outlined in the review section of this manuscript, which are equipped with metering equipment continuously monitoring relevant variables. The constantly collected data has to be converted into a standardized form in accordance with the pre-defined functional unit. It may be necessary to use substitute data, if information for certain production steps cannot be collected or the data quality is insufficient. If so, it has to be clearly stated how and what data was substituted and the data quality shall be outlined in terms of monitoring time, completeness and consistency.

4.3.6 Life Cycle Impact Assessment

Assessed impact categories should be limited to the most relevant impacts and based on quality and/or quantity of flows. Therefore all flows crossing the system boundaries have to be evaluated in terms of their relative amount and their environmental effect/cost prior to elimination [26]. Flows constantly exiting the system during stable production are the carbon fibres itself and treated exhaust gases. Other inputs, which have to be considered, are electric energy, natural gas and nitrogen. As the evaluation of upstream processes is not part of the concept, the effects of the extraction of natural gas and the production of nitrogen, as well as the precursor material are excluded. The basic impact categories to be considered in the core assessment are "climate change (IPCC GWP 100a)", "acidification (CML)" and "eutrophication (CML)".They are based on commonly referenced lists of categories [27] in combination with typically measured compositions of emissions from carbon fibre manufacturing, also mentioned elsewhere [4, 5]. The categories were furthermore chosen based on the most relevant emissions according to environmental regulations and their impact on the respective categories.

4.3.7 Interpretation

To reduce the necessary effort and expert knowledge for interpretation, the assessed impact categories shall be limited to the mentioned categories but can be extended if necessary. To simplify the comparison process a standardized graphical representation of the results, similar to Fig. 4.6 shall be included. Due to mainly relying on directly collected production data and the fact that production runs for specific research purposes cannot be repeated often within the timeframe of a single study, the calculation of uncertainties is not part of the concept. However, as an indicator for deviations the variance of the average values used for calculation should be stated. Analysing hot spots and comparing parameter variations or material changes for various production setups requires standardized results. Therefore the assessment results format shall be predefined and all generated results should be stored in a database for ease of access.

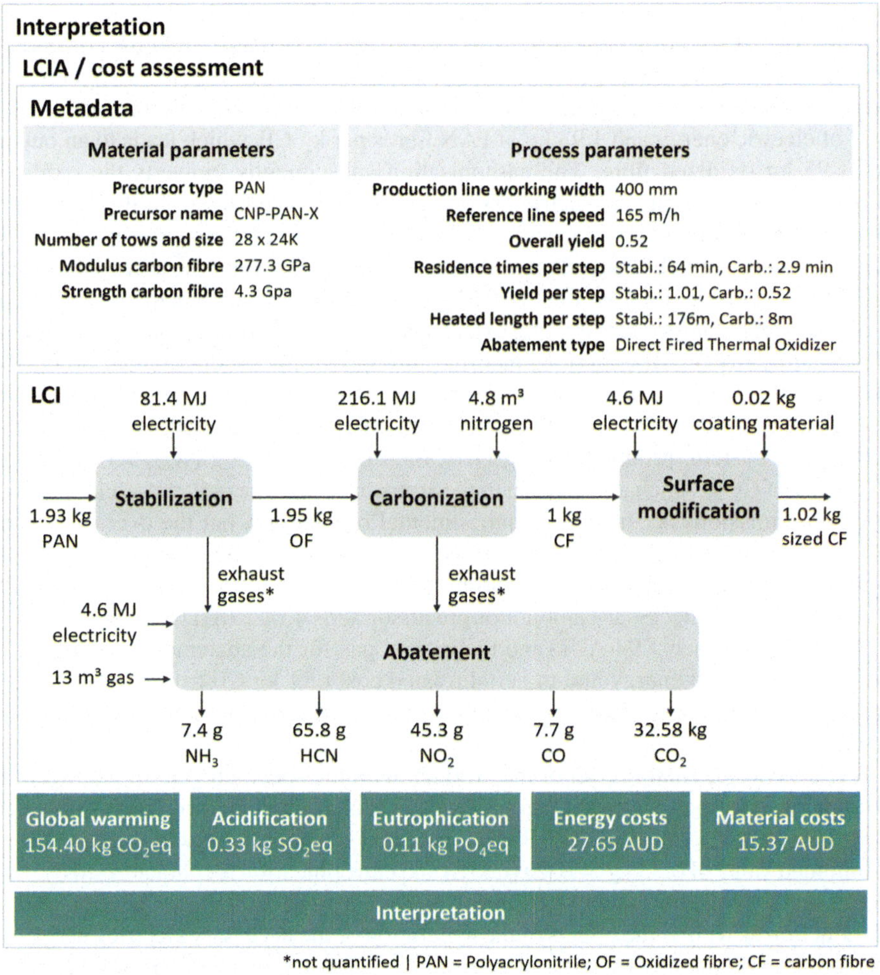

Fig. 4.6 Exemplary application on the Carbon Nexus pilot production line. Reproduced from [3] with permission from Elsevier © Elsevier 2021

4.4 Exemplary Application

The following exemplary application, based on raw data from the Carbon Nexus pilot production line, practically reviews the concept. Scope, system boundaries and functional unit are predefined, as outlined above. Fig. 4.6 summarizes the exemplary results of the assessment. The metadata section illustrates material and process parameters that are required for a fair comparison of the results with other assessments. The LCI section depicts relevant energy and material flows of the considered production and abatement modules. These are summarized to selected environmental and economic impact categories. With these criteria, the stabilization demands 81.4 MJ of electric energy and 1.93 kg of PAN fibres per kg CF, which leads to an output of 1.95 kg oxidized fibre. The carbonisation subsequently converts the oxidised fibres, using 216.1 MJ of electric energy and 4.8 m^3 of nitrogen. In this module 1 kg carbonized fibres are produced. Surface treatment and sizing consume 4.6 MJ electric energy and 0.02 kg of coating material per kg CF. Here the assessed process concludes with the final product, 1.02 kg of sized fibre. The abatement consists of only one dual-stage system treating exhaust gases from all production stages, without any heat recovery. All emissions incurred throughout the production are processed here. As the precise concentrations for each production module are unknown, only 13 m^3 of natural gas and 4.6 MJ of electric energy are specified as input for the abatement system, producing 7.4 g NH_3, 65.8 g HCN, 45.3 g NO_2, 7.7 g CO and 32.58 kg of CO_2, which are released to the atmosphere. These only represent the most relevant emissions according to environmental regulations, but the overall composition of exhaust gases contains additional substances in lower quantities. The total modelled electric energy demand amounts to 85.19 kWh at a cost of 21.29 AUD (0.25 AUD/kWh). The necessary amount of precursor adds 13.22 AUD to the cost, sizing 0.02 AUD, nitrogen 1.94 AUD and the natural gas for the abatement a further 6.36 AUD. The overall energy and material related cost of 1 kg CF from this pilot-scale production setup according to the applied simplified assessment is therefore 43.02 AUD.

The environmental impact of the product, with the specified gate-to-gate view, is summarized in the mentioned categories. The global warming potential is reflected by 154.40 kg CO_2-eq (IPCC GWP 100a), relevant substances for acidification show a combined value of 0.33 kg SO_2-eq (CML) and eutrophication is characterized by 0.11 kg PO_4-eq (CML) for one kg of CF. The values represent the impacts incurred during production as well as the effects of the generation of the necessary electric energy. As the production line the assessment is based on, is located in south-eastern Australia, impacts are calculated with the typical local energy mix. Fig. 4.7 summarizes the environmental impact in the three categories based on direct emissions and emissions embodied in the electricity consumed. It shows that the high electricity demand during production has a major influence on the environmental impact of carbon fibre. With the local energy mix utilized in this study, heavily reliant on fossil fuels, the impact, especially the global warming potential, is quite significant. Global efforts to reduce dependence on fossil energy sources and increasing availability of renewable

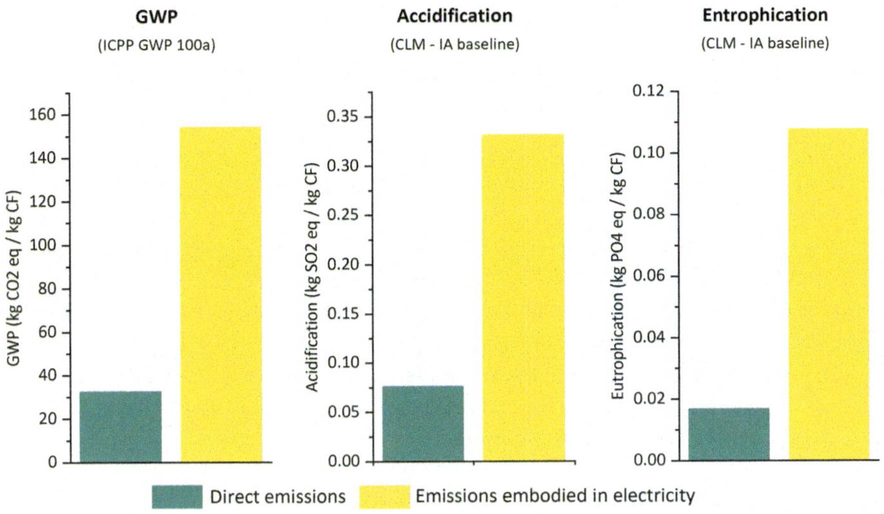

Fig. 4.7 Environmental impact comparison production and electricity generation. Reproduced from [3] with permission from Elsevier © Elsevier 2021

energy can cause a significant improvement in the product's impact. Furthermore, strategic placement of manufacturing facilities in areas with high renewables energy potential, may also yield a benefit. From an Australian perspective, innovative or traditional use of solar energy, and along the coast wind energy, seem to be promising options.

4.5 Production Data Collection, Processing and Analysis

The concept presented in this section heavily relies on data collected during production runs. Therefore the collection, processing, storage and analysis of this data is a crucial part of the project.

4.5.1 Pilot and Research Line Data Collection and Storage

With more than 400 individual parameters on the pilot line and more than 120 on the research line being recorded automatically and stored in a proprietary format on a server, a considerable part of the necessary assessment data can be collected quite efficiently. There are however a few important parameters that are not collected automatically. Furthermore, as the variation in number of recorded parameters suggest, there is a big difference between the research and the pilot line in terms of easily available data. Specifically, there are no energy demands recorded for certain equipment (e.g. electricity for heating elements at research line). In order to record the required additional data, the equipment is fitted with current transducers and a multichannel data logger is used when trials are conducted.

Similarly emissions (composition, concentrations, volume) are not measured constantly. As the exhaust gases, particularly from the low temperature furnace, show very high concentrations of pollution, permanent measurement devices cannot be implemented at reasonable cost. Measurements are therefore taken by hand using a portable FTIR gas analyser in combination with a flowmeter as described in Sect. 3.4.2.

The high number of recorded parameters entails the generation of a significant amount of data for each production run. Therefore it is important that the collected data is efficiently stored and easily accessible.

4.5.2 Data Processing Tool Development

The continuously generated data is stored in a proprietary format on a server but can be manually converted to a comma delimited format (.csv). As the amount of data automatically collected and stored is significant and only a small portion is necessary for the assessment, a (semi-) automated tool is best used to process and assess the data gathered. Based on the overall idea of a simple to use solution which can be accessed without additional software or extensive prior training, an exemplary concept of an assessment tool is implemented in Visual Basic for Applications (VBA) to be used within Excel. The tool automatically processes, analyses and summarizes provided production data. With a growing number of datasets available for individual production scenarios, a database is suggested to be used for ease of storage and efficient use of memory. For this work a SQL database (MySQL) was chosen for initial processing and storage of the production data, as it is common technology to manage large amounts of data. The data was stored on a virtual server and therefore easily accessed through a graphical interface. Development of assessment template.

The data stored in this database, could then be easily accessed through an VBA-based interface in Excel and further processed within the tool to generate the desired results.

In this concept the assessment results were split up into two categories. The first only summarized energy demands and process parameters (Fig. 4.8) and therefore did not require any additional information other than the SCADA data and the basic material and process information shown in Table 4.2. This first part was meant to serve as a quick overview of the production parameters and associated energy demands.

The second part of the tool gave a more detailed overview of the effects, consequences and outputs of a production scenario from an LCA/LCC perspective.

Basic production assessment

Material Parameter	Value	Unit
Precursor type/name	PAN	-
Precursor fibre weight*	3	g/m
Type / Application	Automotive	-
Number of tows*	3	-
Tow size	24	k
Modulus carbon fibre	250	GPa
Strength carbon fibre	4	GPa

Process Parameter	Value	Unit
Production line working width	300	mm
Reference line speed	165	m/h
Overall yield	0.49	-
Oxidation yield	1.01	-
LT yield	0.61	-
HT yield	0.80	-
Residence time Oxidation	64.00	Min.
Residence time LT	1.45	Min.
Residence time HT	1.45	Min.
Heated length Oxidation	176	m
Heated length LT	4	m
Heated length HT	4	m
Abatement type	DFTO	-
Runtime per 1 kg of unsized CF	1.37	h

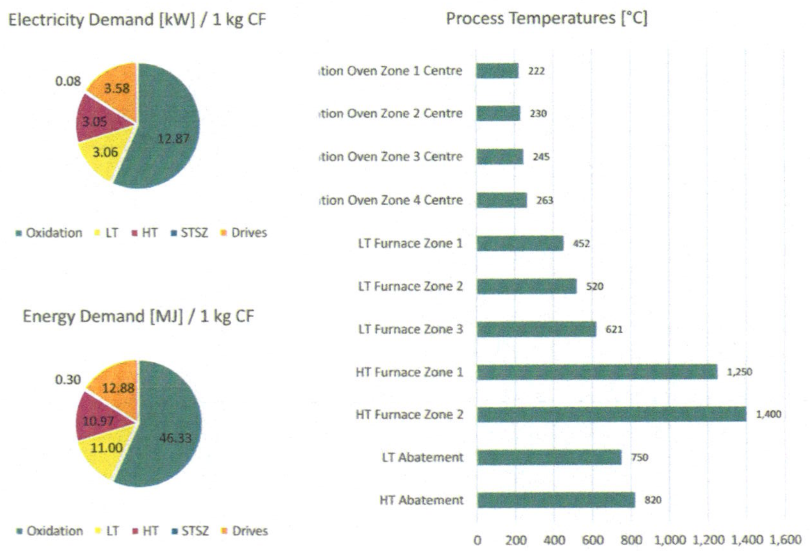

Fig. 4.8 Basic summary assessment results

Table 4.2 Basic assessment tool input parameters

Parameter	Value	Unit
Precursor fibre weight	3	g/m
Number of tows	28	–
Reference line speed	165	m/h
Overall yield	0.49	–
Oxidation yield	1.01	–
LT yield	0.605	–
HT yield	0.8	–
Heated length oxidation	500	m
Heated length LT	5	m
Heated length HT	5	m

Figure 4.9 shows the automatically generated report of the second part of the tool based on information from the first part as well as additional necessary input details as shown in Table 4.3.

The exemplary data shown above was based on the first dataset collected from the Carbon Nexus pilot line for this project in February 2020. The raw data was stored and subsequently processed/analysed during the development of the assessment concept as well as the database and scripts.

Table 4.3 Extended assessment tool input parameters

Emission Parameter	Value	Unit
CO_2	1.4	%
CO	0.9	g/min
HCN	7.7	g/min
NH3	0.87	g/min
NO2	5.3	g/min
Emission temperature	323	°C
Emission flow	18	m/s
Material	Price	Unit
Precursor per kg	5	AUD
Nitrogen per m^3	0.4	AUD
Natural Gas per m^3	0.48	AUD
Electricity per kWh	0.25	AUD
Sizing per kg	13	AUD
Ammonium bicarbonate (ST) per kg	23	AUD

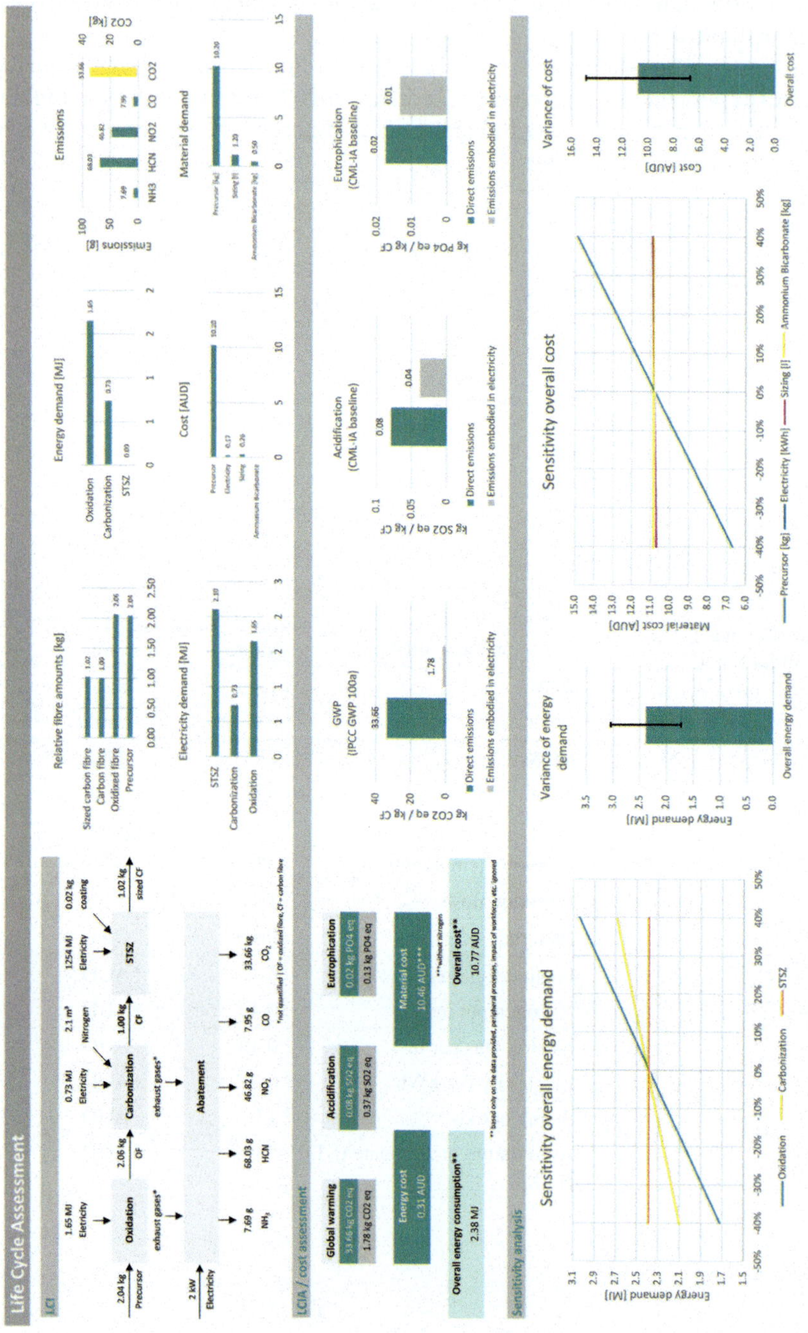

Fig. 4.9 Standardized LCA/LCC results overview

4.6 Conclusion, Critical Review, and Outlook

This chapter presents a modular LCA/LCC-modelling concept to optimize carbon fibre production. It is targeted primarily at research and pilot scale production scenarios. The approach is not intended to substitute fully standard adherent or more extensive assessment methodologies, but rather be integrated in such studies to provide standardized values for comparison. Nevertheless, it can be used as a standalone tool to analyse different scenarios or to get a quick estimate of energy demand, emissions and cost. These streamlined iterative assessments, starting in early research phases, may improve the development of carbon fibres in terms of speed and costs and help focus on environmentally and commercially viable innovations. In an industrial environment, this approach may be incorporated with ongoing economic assessments to achieve a more holistic perspective. As LCA related knowledge often is limited in material focused research areas, a simple-to-use tool developed on the basis of the presented concept, which can be applied without the need of specialized software and intensive prior training, may increase acceptance and adaption. To verify the model, different scenarios, varying process parameters and the used precursor, will be reviewed. This LCA/LCC-based approach could be developed into a more substantial techno-economic-environmental analysis, with the same focus on material development and fine-tuning of production processes. Recycling processes were not considered, as the low amounts of production waste typically are not recycled. In case of industrial mass production, post-production and consumer waste also have to be considered.

References

1. Das S (2011) Life cycle assessment of carbon fiber-reinforced polymer composites. Int J Life Cycle Assess 16(3):268–282. https://doi.org/10.1007/s11367-011-0264-z
2. Baker DA, Rials TG (2013) Recent advances in low-cost carbon fiber manufacture from lignin. J Appl Polym Sci 130(2):713–728. https://doi.org/10.1002/app.39273
3. Groetsch T et al (2021) A modular LCA/LCC-modelling concept for evaluating material and process innovations in carbon fibre manufacturing. Procedia CIRP 98:529–534. https://doi.org/10.1016/j.procir.2021.01.146
4. Griffing E (2009) Carbon fiber from PAN. In: Life cycle inventory summary. Wichita State University
5. Chung DDL (1994) Carbon fiber composites (no. 1). Heinemann, Newton, p 209
6. ISO 14040 (2006) Environmental management—life cycle assessment—principles and framework, I. O. f. Standardization, Geneva, Switzerland. [Online]. Available: https://books.google.com.au/books?id=C_xSygEACAAJ
7. ISO 14044 (2006) Environmental management—life cycle assessment—requirements and guidelines, I. O. f. Standardization, Geneva, Switzerland [Online]. Available: https://books.google.com.au/books?id=1SEkygAACAAJ
8. A. Bjørn, M. Owsianiak, C. Molin, and A. Laurent (2018) Main characteristics of LCA. In: Hauschild MZ, Rosenbaum RK, Olsen SI (eds) Life cycle assessment. Springer International Publishing, Cham, Chapter 2, pp 9–16

9. Hauschild MZ, Herrmann C, Kara S (2017) An integrated framework for life cycle engineering. Procedia CIRP 61:2–9. https://doi.org/10.1016/j.procir.2016.11.257
10. Finkbeiner M, Schau EM, Lehmann A, Traverso M (2010) Towards life cycle sustainability assessment. Sustainability 2(10):3309–3322. https://doi.org/10.3390/su2103309
11. Klöpffer W, Grahl B, Klöpffer W (2014) Life cycle assessment (LCA): a guide to best practice. Wiley, Weinheim
12. Cerdas F, Thiede S, Herrmann C (2018) Integrated computational life cycle engineering—application to the case of electric vehicles. CIRP Ann 67(1):25–28. https://doi.org/10.1016/j.cirp.2018.04.052
13. Herrmann C, Dewulf W, Hauschild M, Kaluza A, Kara S, Skerlos S (2018) Life cycle engineering of lightweight structures. CIRP Ann 67(2):651–672. https://doi.org/10.1016/j.cirp.2018.05.008
14. Ellringmann T, Wilms C, Warnecke M, Seide G, Gries T (2015) Carbon fiber production costing: a modular approach. Text Res J 86(2):178–190. https://doi.org/10.1177/0040517514532161
15. Dér A et al (2021) Modelling and analysis of the energy intensity in polyacrylonitrile (PAN) precursor and carbon fibre manufacturing. J Clean Prod 303:127105. https://doi.org/10.1016/j.jclepro.2021.127105
16. Rödger J-M, Kjær LL, Pagoropoulos A (2018) Life cycle costing: an introduction. In: Hauschild MZ, Rosenbaum RK, Olsen SI (eds) Life cycle assessment. Springer International Publishing, Cham, Chapter 15, pp 373–399
17. Bierer A, Götze U, Meynerts L, Sygulla R (2015) Integrating life cycle costing and life cycle assessment using extended material flow cost accounting. J Clean Prod 108:1289–1301. https://doi.org/10.1016/j.jclepro.2014.08.036
18. Heijungs R, Settanni E, Guinée J (2012) Toward a computational structure for life cycle sustainability analysis: unifying LCA and LCC. Int J Life Cycle Assess 18(9):1722–1733. https://doi.org/10.1007/s11367-012-0461-4
19. Miah JH, Koh SCL, Stone D (2017) A hybridised framework combining integrated methods for environmental life cycle assessment and life cycle costing. J Clean Prod 168:846–866. https://doi.org/10.1016/j.jclepro.2017.08.187
20. Tasala Gradin K (2020) Simplified life cycle assessment approaches and potential impact shifts. School of Industrial Engineering and Management, Kungliga Tekniska högskolan, Stockholm, Sweden
21. Liddell H, Dollinger C, Fischer A, Brueske S, Carpenter A, Cresko J (2017) Bandwidth study on energy use and potential energy saving opportunities in U.S. carbon fiber reinforced polymer manufacturing, USA
22. Arnold U, De Palmenaer A, Brück T, Kuse K (2018) Energy-efficient carbon fiber production with concentrated solar power: process design and techno-economic analysis. Ind Eng Chem Res 57(23):7934–7945. https://doi.org/10.1021/acs.iecr.7b04841
23. Suzuki T, Takahashi J (2005) LCA of lightweight vehicles by using CFRP for mass-produced vehicles. In: Fifteenth international conference on composite materials, Durban, South Africa. Department of Environmental and Ocean Engineering, The University of Tokyo, p 4
24. Al Aiti M, Jehnichen D, Fischer D, Brünig H, Heinrich G (2018) On the morphology and structure formation of carbon fibers from polymer precursor systems. Progr Mater Sci 98:477–551. https://doi.org/10.1016/j.pmatsci.2018.07.004
25. Hur T, Lee J, Ryu J, Kwon E (2005) Simplified LCA and matrix methods in identifying the environmental aspects of a product system. J Environ Manage 75(3):229–237. https://doi.org/10.1016/j.jenvman.2004.11.014
26. Fleischer G, Gerner K, Kunst H, Lichtenvort K, Rebitzer G (2001) A semi-quantitative method for the impact assessment of emissions within a simplified life cycle assessment. Int J Life Cycle Assess 6(3):149–156. https://doi.org/10.1007/bf02978733
27. Stranddorf H, Hoffmann L, Schmidt A (2014) Update on impact categories, normalisation and weighting in LCA. Danish Environmental Protection Agency, Copenhagen, Denmark

Chapter 5 Material Evolution and Emission Analysis of a Pilot-Scale Carbon Fibre Production Process

In many areas the use of composite materials is still in an early phase of development or research but increasing interest highlights their growing importance [2–5]. Today, the vast majority of carbon fibre is manufactured from polyacrylonitrile (PAN) precursor fibres because of the excellent mechanical properties, such as a tensile strength as high as 7 GPa and modulus up to around 400 GPa, low density and excellent chemical stability, except in the presence of oxidizing agents at elevated temperatures [6–9]. They are often categorised by their mechanical properties, such as high modulus (> 300 GPa), intermediate modulus (> 265–300 GPa), standard modulus (230–265 GPa) and high strength (> 3–4 GPa) [10, 11].

5.1 Background

As outlined earlier, precursor fibres are thermally stabilized under tension in air at temperatures between 200 and 300 °C [7, 10] transforming the linear thermoplastic to a crosslinked polymer. The fibres pass through a sequence of ovens at increasing temperatures while controlling tension, gas flow and residence times. The chemical reactions convert the polyacrylonitrile chain structures into a ladder type structure via cyclization, dehydrogenation and oxidation reactions [10, 12]. As this is a complex process, quantifying the chemical changes that occur, has proven to be difficult because the reactions are slow diffusion-driven processes. Additionally, multiple chemical reactions take place simultaneously and make it even more difficult to attribute material changes to specific reactions [13, 14].

Due to the exothermic nature of the stabilization process, the oven conditions are tightly controlled to avoid significant degradation and ensure a high-quality carbon

Chapter reprinted (adapted) with permission from [1]. Copyright 2021 American Chemical Society.

T. Groetsch, *Enhancing Carbon Fibre Production Through Life Cycle Analysis*, Springer Theses, https://doi.org/10.1007/978-3-031-70578-6_5

fibre and yield [15, 16]. Reactive hazardous gases, such as CO, HCN and NH_3 evolve during this stage, requiring the heated air to be carefully circulated and partially exhausted to maintain emissions at concentrations which avoid explosive mixtures with air and maintain stable process conditions [6, 7, 10, 12].

After the stabilization stage, the fibres are heated in furnaces from about 400 °C up to approximately 1600 °C in a two-step carbonization stage [6, 7]. The furnaces are typically purged by a constant flow of N_2 to prevent oxygen from entering and to remove the evolved gaseous emissions. During the earlier stages of carbonization, up to 600 °C, H_2O, CO, CO_2 and NH_3, are primarily evolved. As the temperature increases, the composition of the emissions changes to include H_2, N_2 and HCN [6, 10]. With respect to the evolving micro-structure, up to 600 °C, graphite-like ribbon structures are formed through dehydrogenation causing intermolecular cross-linking [8, 10], but above 600 °C, denitrogenation causes these ribbons to further crosslink into sheet-like structures, creating the turbostratic micro-structure often referred to in literature [7, 8]. During high temperature carbonization, further cross-linking occurs, the distance between the sheets is reduced, and the remaining nitrogen is liberated, leading to an increase in fibre modulus [6, 8, 10]. Finally, during the carbonization stage, fibre density increases and the fibre diameter is considerably reduced [10].

The basic principle of carbon fibre conversion has been in use for decades without significant change, although incremental improvements have certainly been made as manufacturing methods were improved [17] and higher production yields achieved [4]. As the process conditions and detailed production setups are proprietary, there is little detailed information available regarding commercial scale production processes [18]. To foster research towards improvements of the production process and enhance insight into the necessary setup and conditions, it is important to thoroughly understand and detail the production process, potential influence factors and the chemical transformations occurring during processing.

During each process step, gaseous emissions are evolved, with the most significant being HCN and NH_3, as the structure and mass of the fibre material changes [6, 12, 19]. Other process emissions commonly reported include N_2, CH_4, CO, CO_2, C_2H_4, C_2H_6, H_2 and H_2O [6, 10, 19, 20]. Chung [10] based the discussion of emissions from PAN-based carbon fibre production on results reported in the literature. Possible mechanisms for the generation of HCN, NH_3 and H_2 were also outlined. HCN and NH_3 were reported to be the main toxic emissions in this case, there was, however, no further description of the experimental design of research. Figure 5.1 shows select reaction mechanisms and equation thought to generate key emission species.

Morgan [6] summarized the generation of gaseous by-products on a per-production step level, providing an excellent overview of the underpinning chemistry of carbonization. NH_3 and HCN were noted to be the main emissions from the stabilization stage, but other substances such as H_2O, CO_2, CO and tars were also released. The low temperature (LT) furnace produced NH_3, HCN, H_2O, CO_2, CO, CH_4 and high molecular weight compounds at up to 1000 °C. The possible mechanisms responsible for generating the gases were also discussed, some of which are shown in Table 5.1. The high temperature (HT) carbonization part of the production process was briefly discussed and only mentioned HCN as a relevant emission.

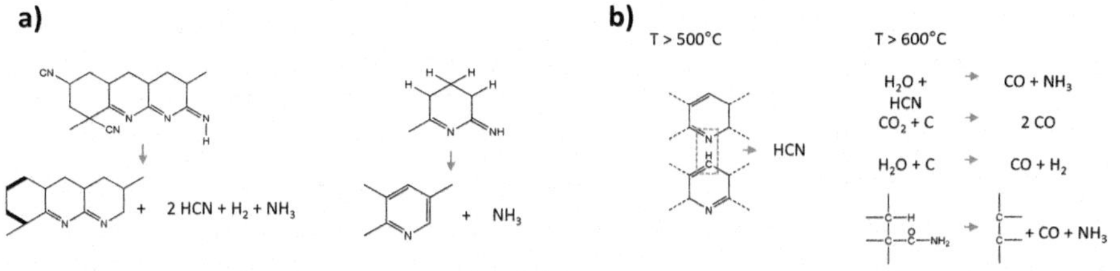

Fig. 5.1 Selected reaction mechanisms and equations generating emissions during **a** stabilization and **b** carbonization

Table 5.1 Evolved gsases and related mechanisms during oxidation and carbonization of PAN fibre

Manufacturing stage	Major emissions	Interpretation	References
Oxidative stabilization < 300 °C	CO, HCN, NH_3, H_2O and CO_2 evolved	HCN removal from depolymerisation,	[2, 21–23]
		NH_3 removal from depolymerisation following cyclization,	
		H_2O emissions from dehydrogenation,	
		CO_2 from –COOH groups in oxidized polymer	
LT carbonization 350 – 1000 °C	Up to 600 °C mainly CO_2, H_2O, CO, and NH_3 evolved, at higher temperatures more N2, HCN and H2 evolution	CO_2 from –COOH groups in oxidized polymer,	[21–24]
		H_2O elimination from intermolecular crosslinking (≈ 400 °C) and dehydrogenation (< 600 °C),	
		HCN elimination from intermolecular crosslinking and depolymerisation (>500 °C),	
		N_2 elimination from denitrogenation and crosslinking (> 600 °C)	
HT carbonization 1000 – 1800 °C	N_2, H_2, and traces of HCN evolved	Cross-linking by denitrogenation	[22–24]

Rahaman et al. [25] reviewed the chemical rseactions occurring during stabilization and carbonization of PAN fibres. HCN, NH_3 and CO were identified as the main toxic emissions along with other volatiles such as H_2O, CO_2 and H_2, with no mention of how the data was derived. In a life-cycle assessment report, Griffing [19] compared values reported in literature with calculations made for a mass balance study. The gases included in the study were HCN, NH_3, CH_4, CO, CO_2, C_2H_4, C_2H_6, H_2, and H_2O, with CH_4, CO, CO_2, C_2H_4, C_2H_6, water and HCN assumed to have evolved at low temperatures while H_2 was only emitted during high temperature carbonization. The study, however, was purely based on mass balance calculations and literature values were used as comparison. Nielsen et al. [26] analysed the formation of gases during the pyrolysis of PAN, focusing on HCN and NH_3 along with H_2O, CO_2 and CO. They suggested that part of the NH_3 was formed from HCN at slow heating rates and also mentioned that substantial amounts of emissions may be contained in tar but did not further analyse the tar. It is important to note that this study was conducted on a laboratory scale. Ko [27] used a small scale production setup to analyse the gases evolved during a continuous stabilization process. Similar to previous examples, NH_3, HCN, H_2O, CO_2, CO were mentioned as emissions, and were treated in incinerators to break down the hazardous components further and comply with environmental regulations.

Most of the studies presented here were either based on results derived using a laboratory bench-top process or did not mention the experimental scale, and often were based on past literature values using old technology. As technology, production equipment and precursors have improved over the years, a detailed investigation of emissions during carbon fibre production utilizing state-of-the-art equipment can provide valuable insight into the current environmental impact while providing new approaches for further improvements. A better understanding of the changes occurring and reactions, as well as the chemical composition of emissions generated during the transformation of the fibre is crucial to enhance overall process efficiency and decrease the environmental impact of carbon fibres. Hence, this study analyses and quantifies generated gases at a scale directly analogous to pilot scale manufacturing of carbon fibre. The resulting carbon fibre analysis is enabling the chemical mechanism to be identified, while relating the changes of the fibre to the measured emissions. Overall, the study correlates the chemical reactions within the transformation of the fibres and generated emissions to the specific stages of the manufacturing process of carbon fibre.

5.2 Materials and Methods

This chapter examines relations between material structure and properties, process conditions and emissions based on what is considered to be representative of current industry standard manufacturing process conditions for a commercially relevant grade of carbon fibres.

Table 5.2 Carbon fibre production process parameters

Stage	Parameters	Values
Feedstock	Set production line speed (m/h)	30
Stabilization (oxidation)	Processing gas	Air
	Flow delivery technology	Forced flow, centre-to-ends
	Number of ovens	4
	Total residence time in heat (min)	48
	Temperature profile per oven (°C)	236 – 246 – 256 - 265
	Tension profile per oven (cN)	3500 – 3300 – 4000 – 4000
Low temp. Carbonization (LT)	Processing gas	Nitrogen
	Number of zones	3
	Total residence time in heat (min)	3
	Temperature profile per zone (°C)	400 – 550 – 750
	Tension (cN)	1500
High temp. Carbonization (HT)	Processing gas	Nitrogen
	Number of zones	2
	Total residence time in heat (min)	2
	Temperature profile per zone (°C)	1200 – 1500
	Tension (cN)	2500

5.2.1 *Manufacture of Carbon Fibre Samples*

Carbon fibres were manufactured using the smaller carbon fibre research line at Carbon Nexus. The study was based on the reference commercial PAN precursor of 24,000 filaments introduced in Sect. 3.2. Three tows of precursor material were simultaneously processed as per specifications given in Table 5.2.

The resultant properties of carbon fibres are shown in Table 5.3.

5.2.2 *Experimental Setup*

The gas evolved during production of the fibre samples were measured in-*situ* at the exhaust stack using a FTIR gas analyser. Figure 5.2 shows the exhaust duct arrangement and FT-IR sampling position labelled on the left. All incinerators or gas

Table 5.3 Carbon fibre material properties

Product	Parameters	Values
Carbon fibres	Elongation at break (%)	1.52
	Tensile strength (GPa)	3.53
	Tensile modulus (GPa)	252.97
	Mass density (g/cm^3)	1.78
	Sizing type	No sizing applied (unsized)
	Material grade	Standard modulus

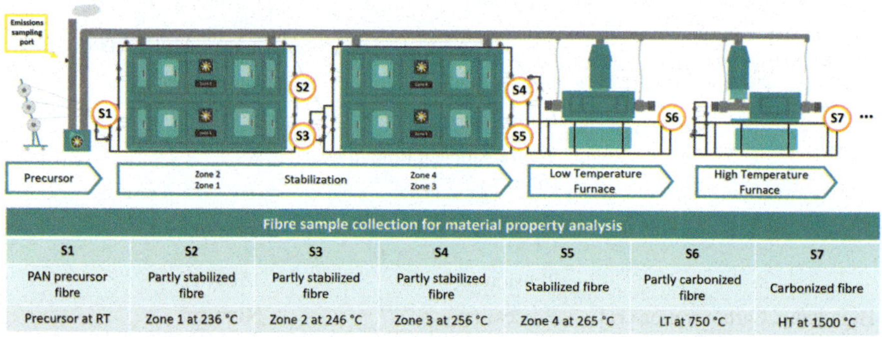

Fig. 5.2 Carbonization line schematic and fibre sample location (orange circled)

treatment devices were temporarily turned off for the purpose of this study. Fibre sampling locations are circled in orange.

The characterization of gaseous emissions was performed using two separate measurement trials (Fig. 5.3). In the first trial (Trial 1), named "full-line", all the gases emitted from the entire manufacturing process were sampled. This configuration provided an overall overview of the gaseous emission from the entire process.

The second experimental trial, named "segmental", comprises the individual characterization of gaseous emissions from the three main manufacturing stages: stabilization, low (LT) and high temperature (HT) carbonization (Trial 2a, b, c). For this setup, each experiment was designed to collect emissions from only one specific stage, therefore the fibres were only threaded through the respective section while

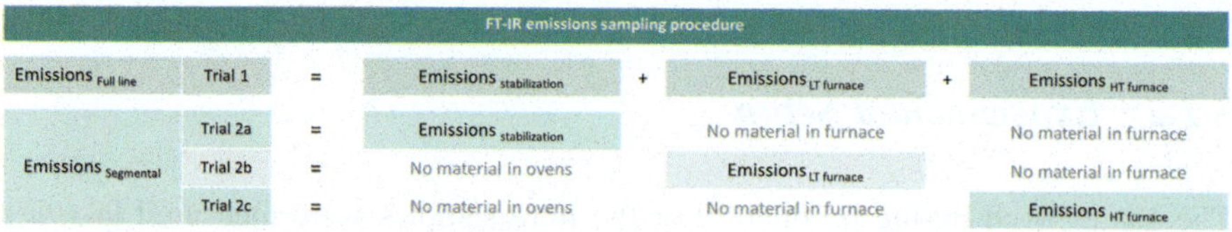

Fig. 5.3 Expersimental configurations

all other process parameters were kept identical. This experimental configuration provided a stage-by-stage gaseous emission analysis of the manufacturing process.

5.2.3 *Gaseous Measurements*

The chemical composition of the evolving gases was determined using a portable Gasmet DX 4000 FT-IR gas analyser (Gasmet Technologies Oy, Finland) equipped with a temperature-controlled sampling unit (APP100). Using the Calcmet software (Gasmet Technologies Oy, Finland), individual gases were identified and the gas concentrations were quantified by comparing the generated infrared absorption spectra against reference libraries of gases of known concentrations [28–30]. The measurement process and subsequent calculations are described in more detail in Sect. 3.4.2. To take the measurements, the sampling probe was inserted into the middle of the exhaust stack duct (ø 150 mm) using the sampling port shown in Fig. 5.2. Each gas sampling experiment was conducted at least 2 times. The gases were sampled at least 15 times for a duration of one minute for each experiment.

The gas flow velocity and temperature of the gases were measured with an anemometer (Kanomax Anemomaster 6162). Measurements were performed for a total duration of 5 min before and after each gas emission measurements.

5.2.4 *Material Characterization*

To complement the gas emission measurements, fibre samples were collected at each step of the production process (Fig. 5.2). The collected material was characterized using various techniques listed below.

5.2.4.1 Mass Density

The mass density of PAN precursor and stabilized polyacrylonitrile fibres were measured at 23 °C using a density gradient column method in accordance with ASTM D1505-10: Standard Test Method for Density of Plastics by the Density-Gradient Technique.

5.2.4.2 Fourier Transform Infrared Spectroscopy (FTIR)

The evolution of the fibre chemistry during the stabilization stage was measured using a Bruker Lumos FTIR spectrometer equipped with a germanium crystal. Each sample was measured at three spots along the fibres with an average of 64 scans per measurement. The evolution of stabilization was quantified using the extent of cyclization reaction (EOR) index (Eq. 8).

5.2.4.3 Elemental Analysis

The CHNS elemental analysis of the fibre samples was performed using a Vario Macro Cube Elemental analyser (Elementar, Germany). All samples were tested at least three times. Prior to the elemental analysis measurement potential contaminants and spin finish were removed by washing the fibres in acetone solutions.

5.2.4.4 Linear Mass Density

Fibre tow samples of 1m length were carefully cut to measure the linear mass density after each manufacturing stage Three fibre sample were tested at each stage.

5.2.5 Tensile Properties

The single filament material properties of intermediate fibres and final carbon fibres were measured on a Favimat + single fibre tensile tester (Textechno, Germany). A minimum of 50 individual filaments were measured for each sample.

5.3 Results and Discussions

Analysis of fibre samples after each manufacturing step throughout stabilization and carbonization enabled an assessment of the gradually evolving chemical composition, while the segmental part of the study provided links to evolving gasses in each section of the process.

5.3.1 The Material Transformation Throughout the Manufacturing Process

The gradually changing chemical composition can be seen in the FT-IR spectra of the fibres throughout stabilization (Fig. 5.4a). The FT-IR spectra of the PAN precursor is characterized by the following absorption peaks: 2939 cm^{-1} (CH- in CH_2 groups), 2242 cm^{-1} (stretching of nitrile functional groups), 1730 cm^{-1} (C=O of the carboxylic groups from acidic comonomers), 1455 cm^{-1} (C-H in CH_2) and 1355

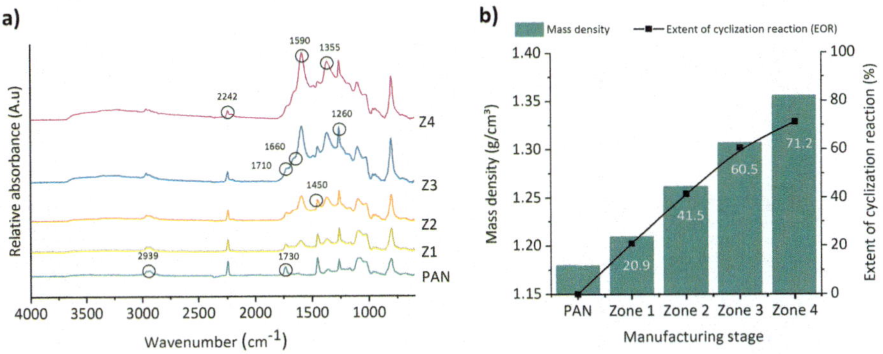

Fig. 5.4 **a** FT-IR spectra evolution of the fibre extracted after each stabilization oven, **b** extent of cyclization reaction and mass density evolution

cm^{-1} (C-H in CH groups) [9, 31]. As stabilization progressed, significant changes to the chemical structure were observed in the FT-IR spectra. The dehydrogenation of the polymer backbone was assessed by the variation in the intensities of the two peaks assigned to C-H stretching. The peak located at 1455 cm^{-1} slowly decreased whereas the peak at 1355 cm^{-1} increased, highlighting the formation of C = CH from CH–CH_2 via the loss of water molecules from the polymer structure [32]. The initiation of cyclization reaction via carboxylic groups of the acidic comonomers was observed by the progressive disappearance of the 1730 cm^{-1} peak [32]. The cyclization reaction was observed by the conversion of pendant nitrile groups (2242 cm^{-1}) to heterocyclic structures (C = C, C = N), indicated by the increase of the 1590 cm^{-1} absorption peak. The intake of oxygen (referred to as oxidation reactions) from these heterocyclic structures was observable with the formation of hydroxyl carbonyl and ketone functional groups (1660 – 1710 cm^{-1} region) which led to a significant increase in the thermal stability of the fibre [31, 32].

As shown in Fig. 5.4b, the extent of cyclization reaction (EOR) increased consistently through stabilization zones 1 to 3 but the reaction rate appears to slow down during zone 4, most likely a result of the decreasing concentration of available unreacted nitrile groups. Importantly, the mass density of the fibre, in the form of the evolution of a compact microstructure was also found to correlate directly with EOR as stabilization progressed [33, 34]. The extent of cyclization after stabilization ultimately reached 71.2%, which is typical of a thermally stabilized carbonization feedstock [16]. The mass density of the stabilized fibre measured, was 1.36 g/cm^3, which is also within the range of mass densities previously reported for high performance carbon fibre production [16, 35].

Furthermore, elemental analysis was carried out on fibres collected after each manufacturing stage (Fig. 5.5). The analysis showed that the PAN fibre had an elemental composition matching previous studies consisting of C, N and H contents of 66%, 25% and 6%, respectively [36–39]. During the stabilization stage, a small

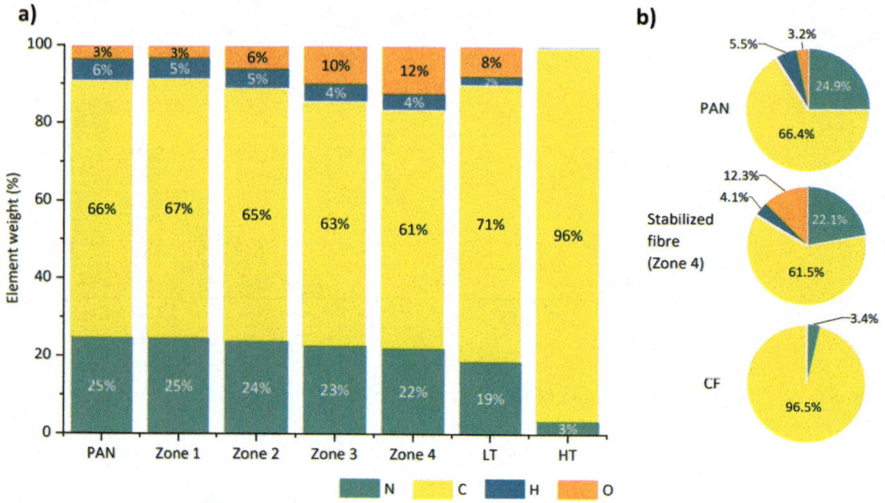

Fig. 5.5 Elemental analysis results for the carbon fibre manufacturing process **a** per process step, **b** for major production stages

decrease in C, H and N content was observed, which has been attributed to the formation of gaseous by-products generated during stabilization [37, 40, 41]. The initial 3% of O from the acidic comonomers increased to approximately 12% by the end of stabilization, complementing the findings of Chung [10]. This is important because oxygen triggers the dehydrogenation step primarily considered to improve thermal stability of the fibre during carbonization [7]. Increasing oxygen can also be directly correlated to an increasing extent of the cyclization reaction index. Johnson et al. [42] previously proposed that dihydropyridine structures are prone to free radical oxidation reactions when exposed to oxygen due to the presence of strongly activated methylene functional groups.

Shown in Fig. 5.6a, the evolution of the fibre mass or linear mass density was studied throughout the manufacturing process. It was observed that the linear mass density decreased in the first stabilization zone, and while the reason for this decrease was unclear, it is proposed to be caused by the partial loss of spin-finish, evaporation of moisture, stretching of the fibre and evolution of gases [6, 10]. Subsequently, the mass density and overall weight increased mostly due to the intake of oxygen molecules already highlighted by the elemental analysis data. This increase was due to the progress of oxidation reactions and was greater than the weight loss arising from by-products formed during the stabilization process [38, 39]. The gas loss from the formation of by-products coupled with the mass gain from the oxidation reactions adds to the complexity to study the chemical transformation of the precursor.

The evolution of mechanical properties throughout the manufacturing process is shown in Table 5.4. During the stabilization stage, the tensile strength and modulus of the precursor fibres decreased from 0.66 GPa and 13.03 GPa to 0.35 GPa and 10.00 GPa, respectively, for stabilized fibres (Zone 4). Concurrently the elongation

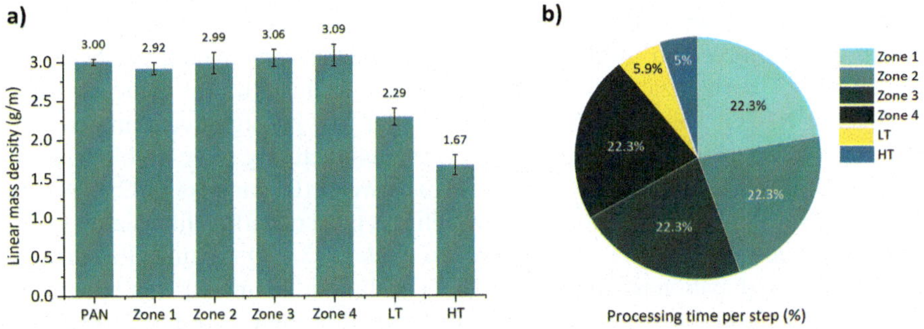

Fig. 5.6 **a** Change of linear mass density throughout production, **b** allocation of processing time per manufacturing step

of the fibre increased from PAN (13.29%) to Zone 4 (16.45%). These changes are thought to be caused by the destruction of PAN crystalline structure and progressing chemical transformation of the PAN precursor into oxidised PAN fibres with a cyclic ladder-like structure [43].

During carbonization, the rate of change in the material property was significantly higher than during stabilization. Indeed, after low temperature carbonization, the weight of the fibre was almost 30% less than the stabilized fibre (Fig. 5.6a). It is important to highlight that this weight loss occurred over a much shorter period of time given the much lower residence time of the fibre in the LT and HT furnaces (Fig. 5.6b). This emphasized the differences between the chemical reaction rates of stabilization and carbonization of the fibre [38, 44].

Previously shown in Fig. 5.5, the carbon content of the fibre increased from 61 to 71% after low temperature carbonization, indicative of the chemical and structural transformation occurring. Interestingly, the elemental composition of fibre after low temperature carbonization still contained significant amounts of oxygen (8 wt%), complementing work reported elsewhere [45, 46]. Between 400 and 600 °C, the hydroxyl groups present in the stabilized PAN underwent crosslinking reactions resulting in the reorganization of the cyclized regions [47, 48]. Intermolecular dehydrogenation occurred, releasing water molecules and first forming planar graphitic

Table 5.4 Mechanical properties of fibre material throughout manufacturing process

Manufacturing stage	Tensile strength (GPa)	Tensile modulus (GPa)	Elongation at break (%)
PAN	0.66 ± 0.01	13.03 ± 0.28	13.29 ± 0.33
Zone 1	0.49 ± 0.03	13.44 ± 0.34	13.86 ± 1.04
Zone 2	0.43 ± 0.03	12.10 ± 0.21	15.52 ± 1.37
Zone 3	0.38 ± 0.04	10.33 ± 0.42	16.34 ± 2.14
Zone 4	0.35 ± 0.03	10.00 ± 0.17	16.45 ± 2.47
LT	1.34 ± 0.15	56.95 ± 1.23	2.45 ± 0.27
HT	3.54 ± 0.73	252.97 ± 9.80	1.52 ± 0.30

layers [47, 48]. Above 700 °C, denitrogenation (N_2) reactions occurred [25, 49], which greatly increased mechanical properties of the carbon fibre produced [46, 50]. After LT carbonization, the fibre processed in this study provided a tensile strength of 1.34 GPa and a modulus of 56.95 GPa, showing significant improvement in mechanical performance from the previous manufacturing stage (Table 5.4).

After high temperature carbonization (HT) between 1000 and 1600 °C, the carbon content of the fibre increased to 96.7%, which is typical of carbon fibres after similar heat treatments [10, 45, 51], as the nitrogen content (3.4%) continued to decrease through denitrogenation reactions. The effect of this carbon increase and the formation of the turbostratic structure ultimately resulted in improved mechanical properties, particularly, tensile strength and modulus which were 3.53 GPa and 252.97 GPa respectively [48]. The mass density of the fibre significantly increased with the progress of crosslinking and aromatization of the heterocyclic rings, as well as the formation of an organized network through repacking of basal planes [45, 52]. During this process, the mass of the fibre was reduced to approximately 55% compared to its initial weight per unit length, which places the overall yield of the carbon fibre production process (without sizing) in the range of reported values [18].

5.3.2 *Gaseous Emissions Occurring Throughout Transformation Process*

To further understand the evolution of the chemical changes and their effect upon micro-structure and property evolution, the exhaust gases were studied using an FT-IR gas analyser. As air is continuously added to the stabilization ovens and nitrogen to the furnaces to replace exhaust gases, the exact input composition of material could not be precisely determined. Therefore, low amounts of gases may be detected that are not a direct by-product of the fibre material conversion process. For this reason this study focused on the analysis of HCN, NH_3, CH_4, CO, which are known to be direct products of the stabilization and carbonization processes.

The overall FT-IR spectra derived from continuous "full-line" processing in Fig. 5.7 includes the relevant reference spectra for each measured gas (reference spectra are not to scale). The HCN reference spectrum shows two peaks in the analysed region between 3400 and 3180 cm^{-1} while the CH_4 reference spectra shows one distinctive peak between 3165 and 2835 cm^{-1}. Similar to HCN, the NH_3 reference spectra displays a distinctive pattern with two peaks between 1150 and 910 cm^{-1} and although a very low level of CO (2200 – 2000 cm^{-1}) was detected, again a distinctive two peak pattern was visible. Good correlation between the measured values of the reference spectrum and actual gases were observed as shown below.

From the spectra the concentration (ppm) for each gas was calculated, which in turn was converted into a mass flow rate in g/h based on Eq. 7. The results obtained for the composition of gaseous emissions are shown in Fig. 5.8, where the results generated per production step are labelled "segmental" to signify the segmental

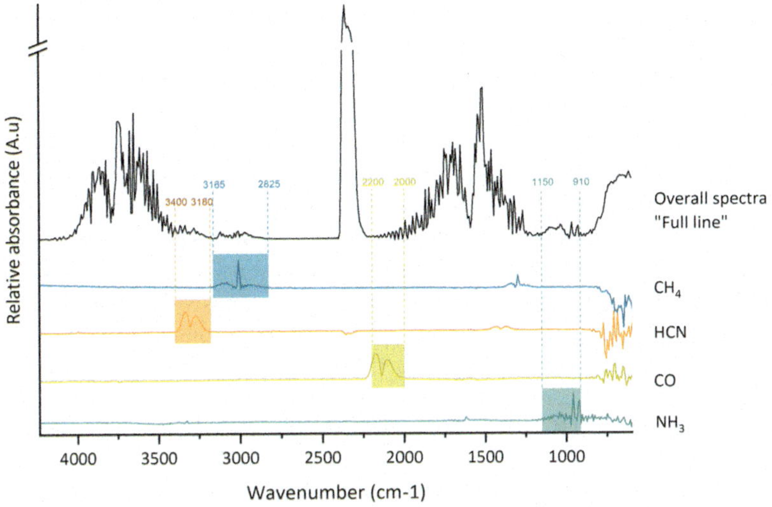

Fig. 5.7 Comparison FT-IR spectrum "full-line" and selected reference spectra

approach, while reference results for the entire production process are labelled "full-line".

Data obtained during the "full-line" line trial showed a gas composition very similar to results previously reported [10, 19]. At 9 g/h, HCN is the predominant gas generated during fibre conversion, followed by approximately 3 g/h of NH_3 and 2 g/h of CO per tow.

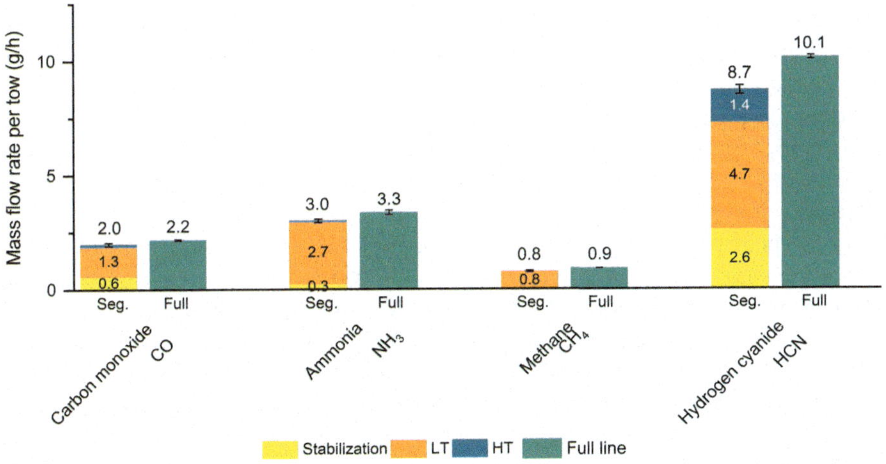

Fig. 5.8 Carbon fibre production emission assessment "segmental" vs. "full-line"

Focusing on the segmental data, the results illustrate the relative contribution of each production stage to the overall emissions and show good correlation with the results from the "full-line" setup. Although HCN was generated in all analysed process steps, the greatest part of the gas was produced in the LT furnace (54%), followed by the stabilization (30%) and finally the HT furnace (16%). Similarly, 90% of the NH_3 was generated in the LT process. A fraction of the gas was released from the stabilization stage (10%) and only insignificant portions were produced in the HT furnace. CO, as the third most relevant gas, was also predominantly produced in the LT furnace (65%), while another 30% and 5% evolved in the stabilization process and HT furnace, respectively. Methane was almost exclusively produced in the LT furnace, with no significant traces of the gas observed during stabilization and high temperature carbonization.

From a manufacturing process perspective therefore, when only the stabilization process is considered, HCN was the dominant gas released arising partially from the fragmentation of the polymer chain (depolymerisation) that does not undergo cyclization [53]. Following HCN, CO was the second largest emission. The loss of carbon-based gases mirrored the elemental analysis data of the fibre showing a slight drop of C content from PAN to zone 4 (Fig. 5.5). The generation of CO was assigned to the partial decomposition and degradation of the oxidising polymer chains and ester units from the acidic comonomers in this temperature range [26, 54]. Finally, a lower concentration of NH_3 was also detected and considered to be the result of the nitrile polymerization from the terminal imine group and progress of the cyclization reactions [36, 55]. Again, good correlation was observed between the elemental analysis data showing a drop of H and N during stabilization to form ammonia-based gases.

For the low temperature carbonization stage (LT), the significant drop of linear mass density, shown in Fig. 5.6a, gave a clear indication where most of the gases were released during the carbon fibre manufacturing process. The emission measurement results for the LT furnace corresponded well with this data, as the LT section was the largest contributor of emissions. HCN was the overall dominant gas and was believed to be a by-product of inter-molecular crosslinking of the stabilized polymer chains [45, 56]. Ammonia was produced in the second highest amount and thought to be released from the termination of active chains or end-to-end joining of ladder structures of stabilized fibres when exposed to high temperatures [25, 56]. Interestingly the production and detection of these two nitrogen-containing gases did not match the elemental analysis data of the stabilized (Zone 4) and LT carbon fibre. While a strong concentration of nitrogen-based gases was observed in the gas FTIR analysis, only a slight drop of nitrogen from the stabilized (Zone 4) to the LT fibre, from 22 to 19%, was observable in the elemental analysis (Fig. 5.5). The reason for this phenomenon is unclear and would require further investigation. The third element emitted in relatively large quantities was CO, followed by CH_4. The loss of these carbon-based gases was assigned to partial degradation and depolymerisation of the chemical structure during rapid carbonization. The presence of oxygen atoms from hydroxyl and carbonyl functional groups resultant from the oxidation reactions during the stabilization may have contributed to the degradation of the chemical structure of the fibre. Indeed, at the stabilization temperatures (235–265 °C), oxygen is highly likely to react with carbon materials [44, 57].

Finally, when the data from the HT furnace was analysed, it was discovered that calculated values underrepresented the emission when compared to the linear mass density. This is an indication that gases other than those investigated in this study were also emitted in moderately high amounts. Other gases likely to be contributing to HT emissions include CO_2, H_2, N_2 and H_2O, as these gases have been detected in previous studies [10, 19, 25, 26]. Furthermore, condensed tars are deposited inside the furnaces and along the exhaust pipes, which also has to be taken into consideration when assessing the results [58]. Findings from the elemental analysis of fibres from each production stage (Fig. 5.5) showed that in the LT furnace the relative amounts of N and O were slightly reduced, while the amount of H was halved. In the HT furnace, the remaining O was fully removed and only a small portion of N remains.

5.3.3 Influence of Ambient Temperature and Humidity

As the chemical conversion processes occurring during the conversion process are driven by molecular reconfiguration, particularly during stabilization, where oxygen and, to a lesser extent, hydrogen play a role, the atmospheric changes in temperature and relative humidity during typical production cycles were briefly examined. While there is some research available on a composite level, as outlined in the review section of this manuscript [59, 60], typical variability and influence on the final products' properties are not well documented. Figure 5.9 shows the typical change in temperature and relative humidity for three manufacturing campaigns throughout the day. Temperatures ranged from a minimum of 6 °C to a maximum of 18°C. The average change throughout each individual day was approximately 7 °C. The relative humidity was observed to be at least 53.6% and reached a maximum of 89.7% with an average change during an individual day of approximately 23%.

The material produced in each trial was analysed and the final results compared. Although the process parameters were not exactly the same each day, the level of mechanical properties for the precursor and final carbon fibres were closely similar. Furthermore no trend was observable that would indicate any dependencies on the atmospheric conditions outside the ovens and furnaces. It is assumed that, in accordance with observations shown in Fig. 5.5 any additional moisture would simply be removed in the first stabilization step and therefore have no further impact on the processes. As even the initial processing temperatures are significantly higher than any expected environmental temperature or changes thereof, no influence on the conversion process is expected. The only influence the temperature may have is the one on the energy demand of heating equipment, where lower temperatures are expected to cause higher energy demand to reach process temperatures. If indeed there is such a correlation for the data collected during the trials conducted in this work, there was no trend observable. The assumption therefore was that the influence is below the margin of general variability of the data.

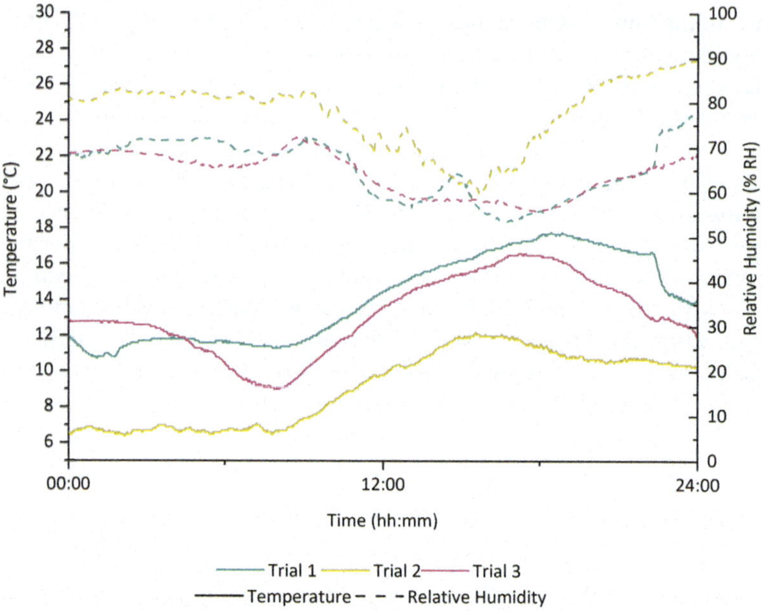

Fig. 5.9 Temperature and relative humidity throughout three trials

5.4 Conclusion

This chapter examined the transformation of polyacrylonitrile (PAN) precursor into carbon fibre, assessed related emissions during the manufacturing stage and linked them to the most commonly reported chemical reaction mechanisms associated with the stabilization and carbonization process. For the first time, this work presented a granular manufacturing step-based approach, which correlated the mechanical and chemical analysis of fibre material for each production step with the evaluation of the segmental emission composition. Each step was analysed in detail and compared with overall results from the process. Material samples and exhaust gases were collected from the stabilization ovens and the LT and HT carbonization furnaces.

The gas FT-IR analysis revealed that HCN (9 g/h) was the predominant gas emitted considering the entire fibre conversion process, followed by NH_3 (3 g/h) and CO (2 g/h). Specifically, 54% of the HCN was generated in the LT furnace, followed by the stabilization ovens and the HT furnace. While 90% of NH_3 was emitted in the LT stage, the remaining 10% was generated in the stabilization stage. Finally, 65% of CO was produced in the LT furnace, followed by 30% emitted during stabilization and a small amount of 5% in the HT furnace. Methane was only detected in the LT furnace in significant amounts.

The emissions generated during the stabilization stage, 2.6 g/h of HCN, 0.3 g/h of NH_3 and 0.6 g/h of CO, corresponded to a reduction of N by 3%, H by 2% and C by 5% according to the elemental analysis of the fibre. The amount of O present in the fibre however increased by approximately 9% in this stage. The high amount of HCN, NH_3, CO and CH_4 measured during LT carbonization, reflected the significant decrease in fibre mass observed during this stage. Elemental analysis also showed a significant drop in element concentrations, except for C. The carbon content continued to increase to 96% during the HT carbonization stage, where all other elements except for 3% N were removed. The related emissions composition showed slightly lower amounts of the four measured gases than expected from elemental analysis results. This was thought to be due to generation of less complex gases and partly conversion of the four measured emissions due to the high temperatures. CO_2, H_2, N_2, H_2O and tar deposits were considered to be potentially contributing to the HT emission composition. The FT-IR gas analysis results obtained for each production step showed good correlation with observations of the entire manufacturing process.

The findings of this chapter foster a better understanding of the correlation between the precursor material, chemical changes and the generated emissions. These insights move research closer to a fully manufacturing data-based mass balance for the stabilization and carbonization process and can be used to uncover potential improvements of the incineration equipment and related gas, heat or energy recovery. Furthermore, the data can be used to compare the impact of different processing conditions or precursor materials and ultimately help steer research towards carbon fibre manufacturing at lower cost.

References

1. Nunna S et al (2017) The effect of thermally induced chemical transformations on the structure and properties of carbon fibre precursors. J Mater Chem A, 5(16):7372–7382. https://doi.org/10.1039/C7TA01022B
2. Gupta A, Kumar A (2010) Composites materials: addressing the climate change. Asia Pac Bus Rev 6(1):78–89. https://doi.org/10.1177/097324701000600107
3. Heuss R, Müller N, van Sintern W, Starke A, Tschiesner A (2012) Lightweight, heavy impact—How carbon fiber and other lightweight materials will develop across industries and specifically in automotive. In: Advanced Industries
4. Koumoulos EP et al (2019) Research and development in carbon fibers and advanced high-performance composites supply chain in Europe: a roadmap for challenges and the industrial uptake. J Compos Sci 3(3):86. https://doi.org/10.3390/jcs3030086
5. Yu M, Wiedmann T, Crawford R, Tait C (2017) The Carbon footprint of Australia's construction sector. Proc Eng 180:211–220. https://doi.org/10.1016/j.proeng.2017.04.180
6. Morgan P (2005) Carbon fibers and their composites. CRC Press, ch. 1174, pp 185–259
7. Frank E, Hermanutz F, Buchmeiser MR (2012) Carbon fibers: precursors, manufacturing, and properties. Macromol Mater Eng 297(6):493–501. https://doi.org/10.1002/mame.201100406
8. Park S-J, Heo G-Y (2015) Precursors and Manufacturing of Carbon Fibers. In: Carbon Fibers S-J, Park (eds) Springer Series in Materials Science. Dordrecht: Springer Netherlands, ch. Chapter 2, pp 31–66
9. Maghe M et al (2016) Using ionic liquids to reduce energy consumption for carbon fibre production. J Mater Chem A 4(42):16619–16626. https://doi.org/10.1039/c6ta06842a
10. Chung DDL (1994) Carbon fiber composites (no. 1). Heinemann, Newton, MA 1994, p. 209
11. Huang X (2009) Fabrication and properties of carbon fibers. Materials 2(4):2369–2403. https://doi.org/10.3390/ma2042369

12. Fitzer E, Frohs W, Heine M (1986) Optimization of stabilization and carbonization treatment of PAN fibres and structural characterization of the resulting carbon fibres. Carbon 24(4):387–395. https://doi.org/10.1016/0008-6223(86)90257-5
13. Edie DD (1998) The effect of processing on the structure and properties of carbon fibers. Carbon 36(4):345–362. https://doi.org/10.1016/s0008-6223(97)00185-1
14. Nunna S, Naebe M, Hameed N, Fox BL, Creighton C (2017) Evolution of radial heterogeneity in polyacrylonitrile fibres during thermal stabilization: an overview. Polym Degrad Stab 136:20–30. https://doi.org/10.1016/j.polymdegradstab.2016.12.007
15. Nunna S, Maghe M, Fakhrhoseini S, Polisetti B, Naebe M (2018) A pathway to reduce energy consumption in the thermal stabilization process of carbon fiber production. Energies 11(5):1145. https://doi.org/10.3390/en11051145
16. Nunna S et al (2019)Time dependent structure and property evolution in fibres during continuous carbon fibre manufacturing. Materials (Basel), 12(7):1069. https://doi.org/10.3390/ma12071069
17. Choi D, Kil H-S, Lee S (2019) Fabrication of low-cost carbon fibers using economical precursors and advanced processing technologies. Carbon 142:610–649. https://doi.org/10.1016/j.carbon.2018.10.028
18. Kaur J, Millington K, Smith S (2016) Producing high-quality precursor polymer and fibers to achieve theoretical strength in carbon fibers: a review. J Appl Polym Sci 133(38):43963. https://doi.org/10.1002/app.43963
19. Griffing E (2009) Carbon fiber from PAN. In: Life cycle inventory summary. Wichita State University
20. Pierson HO (1993) Carbon fibers. In Handbook of carbon, graphite, diamonds and fullerenes: Park Ridge, sNJ, Noyes Publications, pp 166–197
21. Das S (2011) Life cycle assessment of carbon fiber-reinforced polymer composites. Int J Life Cycle Ass 16(3):268–282. https://doi.org/10.1007/s11367-011-0264-z
22. Krautkraemer JA (1998) Nonrenewable Resource Scarcity. J Econom Literat 36(4):2065–2107. [Online]. Available: http://www.jstor.org/stable/2565047
23. Kerns J (2016) Replacing metal with plastic. Mach Des 88(9):34–42
24. Leong WD, Lam HL, Ng WPQ, Lim CH, Tan CP, Ponnambalam SG (2019) Lean and green manufacturing—a review on its applications and impacts. Process Integrat Opt Sustainabil 3(1):5–23. https://doi.org/10.1007/s41660-019-00082-x
25. Rahaman MSA, Ismail AF, Mustafa A (2007) A review of heat treatment on polyacrylonitrile fiber. Polym Degrad Stab 92(8):1421–1432. https://doi.org/10.1016/j.polymdegradstab.2007.03.023
26. Nielsen M, Jurasek P, Hayashi J, Furimsky E (1995) Formation of toxic gases during pyrolysis of polyacrylonitrile and nylons. J Anal Appl Pyrol 35(1):43–51. https://doi.org/10.1016/0165-2370(95)00898-o
27. Ko T-H (1991) Influence of continuous stabilization on the physical properties and microstructure of PAN-based carbon fibers. J Appl Polym Sci 42(7):1949–1957. https://doi.org/10.1002/app.1991.070420719
28. Bacsik Z, Mink J, Keresztury G (2004) FTIR Spectroscopy of the atmosphere. I. Principles and methods. Appl Spectrosc Rev 39(3):295–363. https://doi.org/10.1081/asr-200030192
29. Fu P, Yi W, Bai X, Li Z, Hu S, Xiang J (2011) Effect of temperature on gas composition and char structural features of pyrolyzed agricultural residues. Biores Technol 102(17):8211–8219. https://doi.org/10.1016/j.biortech.2011.05.083
30. Bak J, Clausen S (2002) FTIR emission spectroscopy methods and procedures for real time quantitative gas analysis in industrial environments. Meas Sci Technol 13(2):150–156. https://doi.org/10.1088/0957-0233/13/2/302
31. Nguyen-Thai NU, Hong SC (2013) Structural Evolution of Poly(acrylonitrile-co-itaconic acid) during thermal oxidative stabilization for carbon materials. macromolecules 46(15):5882–5889. https://doi.org/10.1021/ma401003g
32. Ouyang Q, Cheng L, Wang H, Li K (2008) Mechanism and kinetics of the stabilization reactions of itaconic acid-modified polyacrylonitrile. Polym Degrad Stab 93(8):1415–1421. https://doi.org/10.1016/j.polymdegradstab.2008.05.021

33. Bajaj P, Roopanwal AK (1997) Thermal stabilization of acrylic precursors for the production of carbon fibers: an overview. J Macromolecular Sci, Part C: Polymer Rev 37(1):97–147. https://doi.org/10.1080/15321799708014734
34. Ge Y, Fu Z, Zhang M, Zhang H (2020) The role of structural evolution of polyacrylonitrile fibers during thermal oxidative stabilization on mechanical properties. J Appl Polym Sci 138(1):49603. https://doi.org/10.1002/app.49603
35. Takaku A, Hashimoto T, Miyoshi T (1985) Tensile properties of carbon fibers from acrylic fibers stabilized under isothermal conditions. J Appl Polym Sci 30(4):1565–1571. https://doi.org/10.1002/app.1985.070300421
36. Xue TJ, McKinney MA, Wilkie CA (1997) The thermal degradation of polyacrylonitrile. Polym Degrad Stab 58(1–2):193–202. https://doi.org/10.1016/s0141-3910(97)00048-7
37. Kennedy JP, Fontana CM (1959) The thermal degradation of polyacrylonitrile. J Polym Sci 39(135):501–506. https://doi.org/10.1002/pol.1959.1203913542
38. Hanna SB, Yehia AA, Ismail MN, Khalaf AI (2012) Preparation and characterization of carbon fibers from polyacrylonitrile precursors. J Appl Polym Sci 123(4):2074–2083. https://doi.org/10.1002/app.34704
39. Ji M, Wang C, Bai Y, Yu M, Wang Y (2007) Structural evolution of polyacrylonitrile precursor fibers during preoxidation and carbonization. Polym Bull 59(4):527–536
40. Wang PH, Yue ZR, Li RY, Liu J (1995) Aspects on interaction between multistage stabilization of polyacrylonitrile precursor and mechanical properties of carbon fibers. J Appl Polym Sci 56(2):289–300. https://doi.org/10.1002/app.1995.070560221
41. Yu M, Wang C, Bai Y, Wang Y, Wang Q, Liu H (2006) Combined effect of processing parameters on thermal stabilization of PAN fibers. Polym Bull 57(4):525–533. https://doi.org/10.1007/s00289-006-0581-8
42. Johnson JW, Potter W, Rose PG, Scott G (1972) Stabilisation of polyacrylonitrile by oxidative transformation. Br Polym J 4(6):527–540. https://doi.org/10.1002/pi.4980040606
43. Hameed N et al (2016) Structural transformation of polyacrylonitrile fibers during stabilization and low temperature carbonization. Polym Degrad Stab 128:39–45. https://doi.org/10.1016/j.polymdegradstab.2016.02.029
44. Ma Q-S, Gao A-J, Tong Y-J, Zhang Z-G (2016) The densification mechanism of polyacrylonitrile carbon fibers during carbonization. New Carbon Mater 31(5):550–554. https://doi.org/10.1016/s1872-5805(16)60031-8
45. Ko T-H (1991) The influence of pyrolysis on physical properties and microstructure of modified PAN fibers during carbonization. J Appl Polym Sci 43(3):589–600. https://doi.org/10.1002/app.1991.070430321
46. Tse-Hao K, Tzy-Chin D, Jeng-An P, Ming-Fong L (1993) The characterization of PAN-based carbon fibers developed by two-stage continuous carbonization. Carbon 31(5):765–771. https://doi.org/10.1016/0008-6223(93)90013-z
47. Goodhew PJ, Clarke AJ, Bailey JE (1975) A review of the fabrication and properties of carbon fibres. Mater Sci Eng 17(1):3–30. https://doi.org/10.1016/0025-5416(75)90026-9
48. Donnet J-B, Chand Bansal R (1998) Carbon fibers. CRC Press. ch. 572, pp 1–370
49. Zhu D, Xu C, Nakura N, Matsuo M (2002) Study of carbon films from PAN/VGCF composites by gelation/crystallization from solution. Carbon 40(3):363–373. https://doi.org/10.1016/s0008-6223(01)00116-6
50. Laffont L, Monthioux M, Serin V, Mathur RB, Guimon C, Guimon MF (2004) An EELS study of the structural and chemical transformation of PAN polymer to solid carbon. Carbon 42(12–13):2485–2494. https://doi.org/10.1016/j.carbon.2004.04.043
51. Frank E, Steudle LM, Ingildeev D, Sporl JM, Buchmeiser MR (2014) Carbon fibers: precursor systems, processing, structure, and properties. Angew Chem Int Ed 53(21):5262–5298. https://doi.org/10.1002/anie.201306129
52. Ko T-H, Huang L-C (1998) The influence of cobaltous chloride modification on physical properties and microstructure of modified PAN fiber during carbonization. J Appl Polym Sci 70(12):2409–2415. https://doi.org/10.1002/(sici)1097-4628(19981219)70:12%3c2409::Aid-app13%3e3.0.Co;2-0

53. Gupta AK, Paliwal DK, Bajaj P (1996) Effect of the nature and mole fraction of acidic comonomer on the stabilization of polyacrylonitrile. J Appl Polym Sci 59(12):1819–1826. https://doi.org/10.1002/(sici)1097-4628(19960321)59:12%3c1819::Aid-app2%3e3.0.Co;2-t
54. Grassie N, McGuchan R (1971) Pyrolysis of polyacrylonitrile and related polymers—III. Thermal analysis of preheated polymers. Eur Polymer J 7(10):1357–1371. https://doi.org/10.1016/0014-3057(71)90032-2
55. Turner WN, Johnson FC (1969) The pyrolysis of acrylic fiber in inert atmosphere. I. Reactions up to 400°C. J Appl Polym Sci 13(10):2073–2084. https://doi.org/10.1002/app.1969.070131005
56. Watt W (1985) Chemistry and physics of the conversion of polyacrylonitrile fibres into high-modulus carbon fibres. In: Watt W, Perov B (eds) Handbook of Composites, vol. 1,: Elsevier Science Publishers B. V., pp 327–387
57. Yang J, Liu J, Liu W, Wang J, Tang T (2015) Recycling of carbon fibre reinforced epoxy resin composites under various oxygen concentrations in nitrogen–oxygen atmosphere. J Anal Appl Pyrol 112:253–261. https://doi.org/10.1016/j.jaap.2015.01.017
58. Pham HH, Lim Y-I, Ngo SI, Bang Y-H (2019) Computational fluid dynamics and tar formation in a low-temperature carbonization furnace for the production of carbon fibers. J Ind Eng Chem 73:286–296. https://doi.org/10.1016/j.jiec.2019.01.040
59. Dhieb H, Buijnsters JG, Elleuch K, Celis JP (2016) Effect of relative humidity and full immersion in water on friction, wear and debonding of unidirectional carbon fiber reinforced epoxy under reciprocating sliding. Compos B Eng 88:240–252. https://doi.org/10.1016/j.compositesb.2015.11.011
60. Wang S, Kowalik DP, Chung DDL (2010) Effects of the temperature, humidity, and stress on the interlaminar interface of carbon fiber polymer-matrix composites, studied by contact electrical resistivity measurement. J Adhes 78(2):189–200. https://doi.org/10.1080/00218460210384

Chapter 6
Environmental, Property and Cost Impact Analysis of Carbon Fibre at Increasing Rates of Production

While CFRP offer a significant weight reduction of up to 60% compared to standard steel, high tensile strength and modulus and corrosion resistance, the widespread use of CFRP however, has been limited due to one key factor; its production is significantly more energy intensive than that of a comparable amount of steel [2, 3]. Furthermore, approximately half of the initial mass of the precursor fibre is lost as volatiles during the production process, emitting carbon monoxide, carbon dioxide, hydrogen cyanide and ammonia as shown in the previous chapter [4]. Despite the key hazardous substances being incinerated, the overall amount of greenhouse gas emissions continues to increase [4, 5]. Research towards overall improvement of the production process and the reduction of material losses, i.e. emission of off-gases, is therefore an important task for a more widespread application of carbon fibres.

6.1 Background

Although the cost of carbon fibre has decreased and production volume increased considerably over the years, the basic manufacturing process has remained largely unchanged. Slow thermal stabilization is followed by carbonization of the oxidized precursor. The much slower stabilisation process compared with carbonisation arises from the poor heat tolerance of the precursor fibre and the need to control the exothermic reaction during oxidation. This means that the stabilisation process is highly energy intensive, leading to intensive research dedicated to increasing the rate of oxidation. No doubt manufacturing processes and the utilized equipment have been continuously and incrementally improved over the years, but the secrecy of commercial carbon fibre production has inhibited independent research on minimising energy

This chapter (modified) was published in [1], Copyright Elsevier 2023.

T. Groetsch, *Enhancing Carbon Fibre Production Through Life Cycle Analysis*, Springer Theses, https://doi.org/10.1007/978-3-031-70578-6_6

consumption and environmental impact of carbon fibre manufacture [6]. Consequently, there is a need to understand how optimizing process parameters can reduce energy demand, cost and emissions, while improving material properties on a continuous production line that is representative of commercial setups. Unfortunately, the few studies that focus on the energy demand of a continuous carbon fibre manufacturing process, often lack important details and comparability is low due to the lack of standardised procedural definition, as outlined in Sect. 2.5.

This chapter presents an in-depth understanding of the impact of process parameters in individual production steps on energy consumption and environment to provide new insights into process optimization and reducing the cost and environmental impact of carbon fibre. The high energy demand and low yield is one of the key factors driving the high cost and inhibiting a more widespread application of CFRPs. Evaluating the manufacturing process on a holistic level and understanding key contributors and influential parameters is thus crucial to an efficient and economically competitive production. This chapter describes the manufacture of carbon fibres at increasing rates and temperature profiles, and the material properties, energy demand and emission composition were studied. This was done by systematically increasing the temperatures in the oxidation ovens and the low temperature furnace while reducing residence times through increasing line speed. The temperature profile of the high temperature furnace however, remained the same for each trial. A key goal was to manufacture carbon fibre with similar mechanical properties to enable clearer comparison of the higher production rates in terms of energy demand, emissions, yield and cost.

6.2 Materials and Methods

The work presented here focuses on three key areas to compare four scenarios with an increasing rate of manufacture: material properties, energy demand and emission composition. The parameters referred to for the material properties are, as in previous chapters, the commonly referenced tensile strength, modulus and mass density. The energy demand analysis focuses on the heating elements in the ovens and furnaces as the main contributors, in line with the simplified approach. The emission composition analysis is based on the same principles as outlined in the previous chapter and includes HCN, NH3, CH4 and CO.

To ensure the data collected from the trials could be compared effectively, the study used the system boundaries and simplification strategy of the assessment framework proposed in Chap. 4. The production process as described earlier, was divided into three major sections: thermal stabilization, carbonization and surface finishing in line with the work of previous chapters.

6.2.1 *Material Properties*

All material related experiments in this section were performed using a commercial 24,000 filament (24 k) PAN precursor, the same material used for work in previous chapters. The chemical structure of the source material includes acrylonitrile (AN), methyl acrylate (MA), itaconic acid (IA) with a silicon oil-based spin finish applied for improved handling. The precursor material was processed under conditions akin to commercial production, when scaled to size, with material parameters commonly referenced as commercially relevant grade of carbon fibres [7, 8]. These properties were chosen based on typical properties of commercially available high quality, low-cost carbon fibres intended for automotive or wind turbine applications [9, 10]. The targeted values for the main material properties were a tensile strength of 3.6 ± 0.3 GPa, an initial tensile modulus of 245 ± 10 GPa and mass density in the range of 1.7–1.9 g/cm^3.

6.2.2 *Carbon Fibre Manufacture*

For this chapter, data was collected from the research line at Carbon Nexus, Deakin University (Australia), a scaled down version of a typical commercial production setup with the specifications shown in Table 2.1.

In line with previous work, the energy demand was calculated based on SCADA data and, where no data was available, additional current transducers. Emissions were measured using Fourier Transform Infrared Spectroscopy (FT-IR) gas analysis, while material characteristics after each manufacturing step were monitored using single fibre tensile testing, density and FT-IR spectroscopy.

6.2.3 *Processing Conditions*

Four process scenarios with varying processing speeds (Slow, Medium, High, Fast) were investigated. Table 6.1 summarizes the detailed processing conditions including variations in temperature and tension profiles for each experimental layout. The conditions were chosen in a way to keep the overall temperature level as low as possible, as this was thought to be the most energy efficient way of designing the study. The HT carbonization zones therefore were set to the lowest possible fixed temperatures between 1100 and 1300 °C. When increasing the material throughput, adjustments to processing temperature need to be made to assure adequate progress in chemical reactions during the stabilization and carbonisation steps, which in turn increases the energy demand. To achieve the desired level of mechanical properties at increasing processing speeds, the temperature was systematically increased in each stabilisation oven and the LT furnace zones. Mass density and EOR were used

Table 6.1 Process parameter sets for the four processing conditions examined

Stage	Parameters	Values			
	Reference name	Slow (standard)	Medium	High	Fast
Stabilization (oxidation)	Total time in heat (min)	72	58	48	36
	Temperature profile per oven (°C)	229–240 to 249–259	236–245 to 255–264	240–249 to 259–268	244–253 to 262–271
	Tension profile per oven (N)	35–35 to 38–38	33–34 to 37–38	32–33 to 40–40	37–41 to 40–40
Low temp. carbonization (LT)	Total time in heat (min)	4.5	3.6	3	2.25
	Temperature profile per furnace (°C)	400–550 to 750	400–600 to 800	450–750 to 900	450–750 to 950
	Tension (N)	26	25	28	23
High temp. carbonization (HT)	Total time in heat (min)	3	2.4	2	1.5
	Temperature profile per furnace (°C)	1100–1300	1100–1300	1100–1300	1100–1300
	Tension (N)	36	35	35	35

to optimize the stabilization processing conditions and achieve the desired material properties by keeping both variables in a close range for all four trials.

6.2.4 Assessment Approach

The assessment methodology utilized to compare the four production scenarios is described in detail in chapter 4. As the methodology only focuses on directly contributing key areas, peripheral processes are again not part of the assessment as their contributions were deemed negligible or static for the purpose of this work. Although a significant contributor to energy demand and in line of flows pertaining to the environmental impacts, the incineration technology used in the current production environment is not fully representative of commonly used industrial processes. Emissions were therefore analysed before any treatment and hence not combined in environmental impact categories as these emissions would undergo further/different treatment in an industrial setting before release into the atmosphere.

The reference flow is again defined to be the amount of precursor fibre required in each step to produce 1 kg of unsized carbon fibre. The key data for manufacturing trial analysis is collected directly from the equipment and material, as outlined in the following sections.

6.2.5 Energy Measurement

To calculate the real time power demand/current for consumers such as circulation fans or drive motors, Eq. 1 (three phase) and 2 (single phase) were used. As recorded data showed no significant fluctuation in recorded power or current during stable state production, the energy demand was simply calculated using Eq. 3. As not recorded by the SCADA system, data from the heating elements of the ovens and furnaces was gathered using Fastron AT 150 B10 self-powered split core AC current sensors with average RMS output (0 – 10V) and a Hioki Memory HiCorder MR8875 at a rate of 1Hz. Energy consumption was subsequently calculated using the formula mentioned above.

6.2.6 Gaseous Measurements

To analyse the composition of emitted gases again a portable Gasmet DX 4000 FT-IR gas analyser (Gasmet Technologies Oy, Finland) was used. The infrared absorption spectra generated from the emission mixture were compared against reference libraries, as outlined in the previous chapter.

Velocity and temperature were again measured for 5 min before and after each spectroscopic with an anemometer (Kanomax Anemomaster 6162). Analysis in this chapter again focused on four gases, HCN, NH3, CH4 and CO, based on the amount produced during the conversion process and their toxicity. CO2 could not be included as the system is open and the gas is present in the atmosphere preventing accurate quantitation.

6.2.7 Material Characterization

Fibre material was collected after each stage of the manufacturing process. The same techniques and equipment with the same settings used in chapter 5 were used to characterize the samples.

6.3 Results and Discussion

The overarching goal of this chapter was to generate a holistic evaluation of links between the rate of manufacture, energy demand, emissions and the material transformation. The approach was to manufacture carbon fibres with consistent material properties (≈3.6GPa tensile strength, ≈245GPa initial tensile modulus) under varying processing conditions outlined in Table 6.1 and perform an energy and cost analysis for the different pathways. The primary aim was to investigate whether faster production led to lower energy demand per kg carbon fibre and to what degree increasing production speed affects yield and the amount of gases emitted.

6.3.1 Process Optimization

Through optimization of the process parameters, temperature–time profiles were designed for four different production rates: “Slow”, “Medium”, “High” and “Fast”. During the stabilization stage, the temperature was increased by approximately 5 °C, for each zone with increasing line speed (Fig. 6.1a), while the processing time was reduced by 19%, 33% and 50% for the “Medium”, “High” and “Fast” trial. The LT carbonization temperatures ranged from 400–550 °C (Zone1) to 750–950 °C (Zone3) for the four trials as shown in Fig. 6.1b. As mentioned in Sect. 2.4 the HT carbonization zones were selected at the lowest possible fixed temperatures. Figure 6.1c shows that even though the HT furnace temperature was the same for all four trials, the heating rate from 1100 to 1300 °C for the material significantly increased as the line speed increased and residence time decreased.

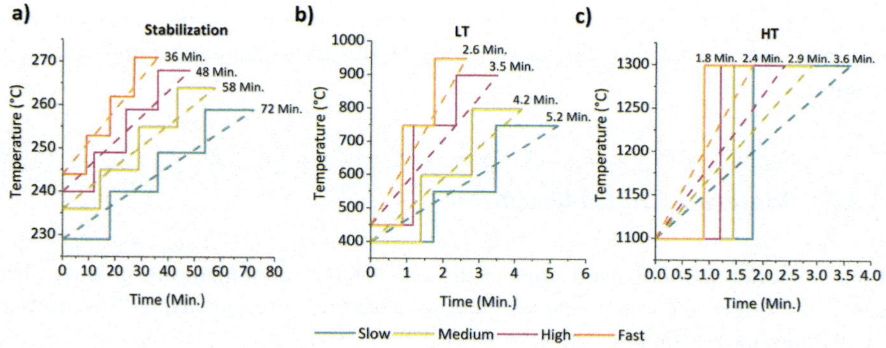

Fig. 6.1 Processing speed and temperature gradient for **a** stabilization, **b** LT and **c** HT for the four processing conditions examined. Reproduced from [1] with permission from Elsevier © Elsevier 2023

6.4 Evolution of Fibre Properties as a Function of Processing Speed

The evolution of material properties throughout the conversion process was studied using various techniques to generate commonly referenced material properties for comparison.

6.4.1 *Extent of Cyclization Reaction and Fibre Mass Density*

Fibre samples were taken after each stage of the manufacturing process and the progress in chemical transformation and evolution of fibre properties was analysed. As shown in Fig. 6.2a, the extent of cyclization reaction (EOR), indicator of one of the most significant chemical reactions occurring during stabilization, increased steadily, but slowed down in the last oxidation zone, likely as a result of less available unreacted nitrile groups. The rate of reaction varied according to the temperature profiles [11, 12]. The densification of the fibre, a result of the development of an increasingly packed microstructure and oxygen uptake by the fibres, directly correlate with EOR throughout the stabilization process [13–15]. The extent of cyclization was calculated to be 73.6 ± 1.8% after stabilization, which is in the range of what is typically targeted for thermally treated precursor material [13]. The stabilized fibre mass density reached 1.37 ± 0.01 g/cm3, which is in the same range as previously reported values (Fig. 6.2b) [13, 16].

The occurring oxidation reactions were more significant than the generation of by-products formed during the stabilization process and therefore the mass density increased [17, 18]. Indeed, after stabilization, during carbonisation, the mass density

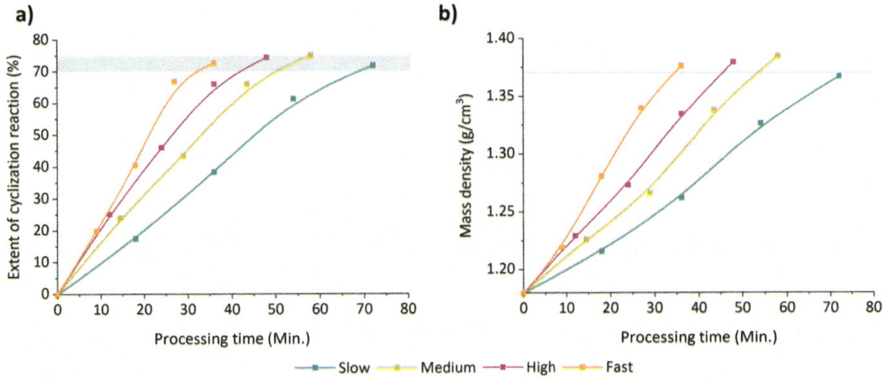

Fig. 6.2 Comparison of **a** extent of cyclization reaction and **b** mass density throughout stabilization section of the manufacturing process. Reproduced from [1] with permission from Elsevier © Elsevier 2023

of the fibre continues to increase as crosslinking and aromatization of the heterocyclic rings progress and basal planes are repacked to form an organized network [19, 20].

Three samples, 1m long, were taken at each stage of the manufacturing process to measure the linear density, showing a decrease in the first stabilization zone, possibly caused by partial removal of spin-finish and moisture as well as stretching of the fibre material (Fig. 6.3) [5, 21]. After the initial decrease, the linear density steadily increased throughout stabilization due to intake of molecular oxygen and reorganization of the microstructure, before the progressive removal of non-carbon material during carbonization caused a sharp decrease [5]. Importantly, the “Slow”, “Medium” and “High” processes displayed similar carbon yields of 54.0, 57.4, and 53.6% as shown in Fig. 6.3, which is in line with reported values for carbon fibre production (without sizing) [22]. This is despite the significantly more aggressive treatments of the fibres and the shorter residence times. However, the “Fast” trial diverged more noticeably, displaying a significantly lower yield of only 47% and a higher variance, showing the limits of increasing the temperatures and process speeds.

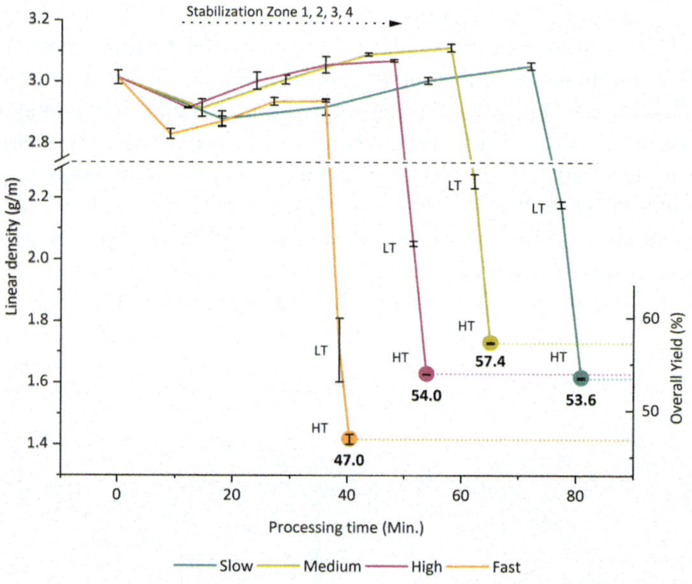

Fig. 6.3 Change of linear mass density throughout process and overlaid final yield of CF correlated to the linear mass density of the final carbon fibre. Reproduced from [1] with permission from Elsevier © Elsevier 2023

6.4.2 Mechanical Fibre Properties

Figure 6.4 shows how the mechanical properties of the material change throughout the conversion process. Tensile strength (0.49 GPa) and modulus (13.43 GPa) of the original precursor fibre decreased by an average of 28.6% and 23.1%, respectively, during stabilization. The reduction in mechanical properties was thought to be a consequence of the partial restructuration of the precursor crystalline structure and the chemical conversion of the PAN into oxidised material with an extensive cyclic ladder structure [23].

After low temperature carbonization, the tensile strength increased by an average of 322.4% to 1.58 GPa compared to the PAN precursor. Similarly, the modulus increased by 718.2% to an average of 96.45 GPa. The evolving mechanical properties are the result of crosslinking reactions of the hydroxyl groups between 400 and 600 °C, which results in the reorganization of the cyclized structure [24, 25]. Furthermore, intermolecular dehydrogenation releases water molecules and the first planar graphitic layers are formed [24, 25]. Above 700 °C, denitrogenation reactions significantly increase mechanical properties through further crosslinking of these planar structures into more continuous sheetlike layers [21, 26–28]. During high temperature carbonization (HT) between 1000 °C and 1300 °C, the formation of turbostratic structures further improved mechanical properties [25], resulting in a tensile strength between 3.22 and 3.84 GPa and a modulus between 242.03 and 254.22 GPa. With the increasing changes in material properties during low temperature carbonization, differences between the four residence times became more pronounced and the material visually was less uniform especially observed in the "Fast" trial.

Figure 6.5 shows individual single filament tensile test results overlayed with a 95% confidence ellipse for the four manufacturing rates, depicting breaking strength vs. modulus and strength vs. maximum elongation. As can be seen, there is no

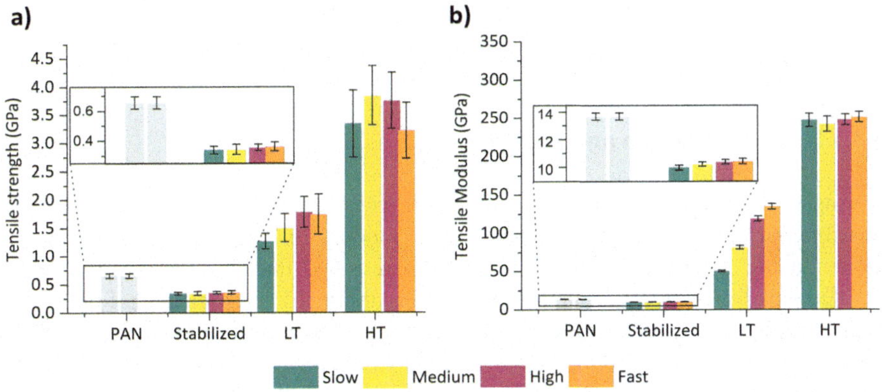

Fig. 6.4 Comparison of evolving material properties **a** tensile strength **b** tensile modulus, at key process stages for the different processing conditions. Reproduced from [1] with permission from Elsevier © Elsevier 2023

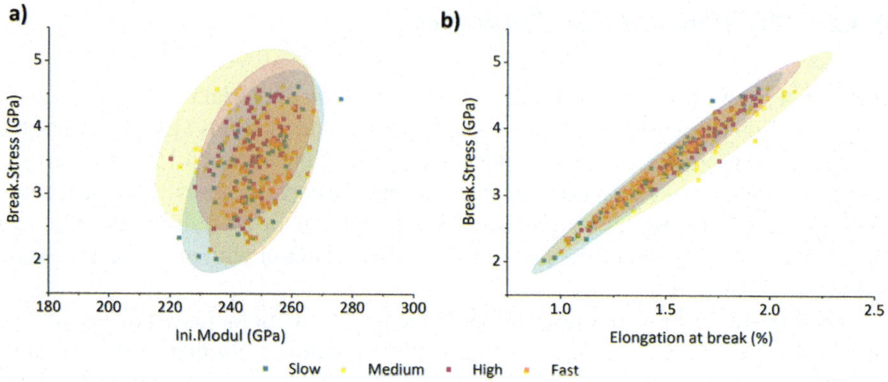

Fig. 6.5 Comparison of the spread of three key material properties, tensile strength, modulus, elongation at break, for carbon fibres produced under the four processing conditions of this study. Reproduced from [1] with permission from Elsevier © Elsevier 2023

significant variation in the spread of single test results, further confirming the minimal impact upon fibre properties as a result of varying production rate. Each dataset shows a similar x/y-spread and the sizes of the confidence ellipses are also similar with slightly varying mean values between the trials.

6.4.3 Process Emission and Cost Analysis

Two of the most significant drawbacks of carbon fibre-based components or products is the significant environmental impact stemming from the fibre production and the high cost of the material. As much research is committed to improving either one of these or both, a closer look at the impact of process parameter variation on emission composition and cost can help steer research towards promising improvement routes.

6.4.3.1 Gas Evolution and Emission

During each of the four trials, gas emissions were captured and analysed to assess possible variations in either composition or amount of evolving gases as a result of increasing processing speeds. The results revealed that the composition of the emitted gases was very similar complementing the yield and mechanical properties results discussed above. Figure 6.6 displays the emission of analysed gases (g/kg CF) during the entire production for all production speeds. At an average of 194g, HCN was the main gas emitted during fibre conversion, followed by approximately 63g of NH3 and 44g of CO per kg carbon fibre produced, while only 20g of CH4 was emitted on average. Interestingly, the three slower manufacturing rates (longer residence times) displayed very similar emissions for HCN, NH3, CH4, less than

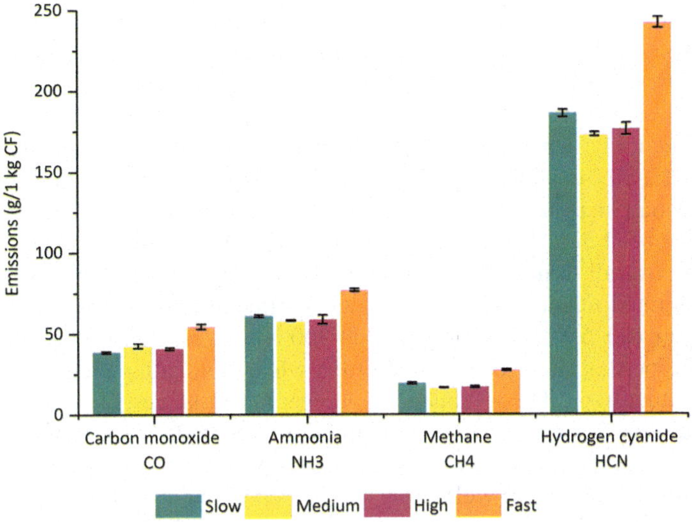

Fig. 6.6 Normalized, gaseous emissions per kg carbon fibre for the four procession conditions. Reproduced from [1] with permission from Elsevier © Elsevier 2023

5% difference being observed after normalisation to the same amount of carbon fibre produced. The "Fast" trial however showed significantly higher concentration of all measured gases, with the highest value for HCN at 242g, which is linked to the increased mass loss and linear density for the"Fast" manufacturing process. The slower three tests showed that the emitted gas concentration only depended on the amount of material being processed and was not affected by other process parameters. The "Fast" trial however suggests that this is only true in a limited temperature and processing time range.

6.4.3.2 Energy Demand Analysis

The energy demand was continuously measured for at least 1 h for each trial after the process had reached a stable state. Applying Eq. (1.3) in combination with the calculated yield for each trial, the energy demand to produce 1 kg of carbon fibre was calculated. Figure 6.7 shows the summarized results with the highest overall energy demand for the "Slow" trial. For the "Medium" trial, a decrease of necessary energy of $19 \pm 1\%$ for each stabilization zone was observed, a 11% decrease for the LT as well as 24% less energy for the HT despite the shorter residence time. At "High" production speed, the analysis showed a $32 \pm 1\%$ reduction in energy used compared to the initial "Slow" trial for each stabilization zone. Interestingly, the energy demand only decreased by approximately 6% in the LT furnace compared to the "Slow" trial. Higher temperatures in all LT zones caused a rise in energy demand, overriding the effect of the shorter residence time. The HT, with temperatures also the same for the

"High" trial, showed a 36% lower energy demand due to the further decreased residence time. Overall, a clear trend was visible for the stabilization section such that decreasing residence time, i.e. higher production speed, led to a decreased overall energy demand per material output, with the exemption of "Fast" production speed. This is particularly important since stabilization is considered to account for approximately 62% of the overall energy demand to manufacture carbon fibres [29, 30]. The relatively small increase in the stabilization temperature profile of 5 °C with varying production speed did not cause a significant enough increase in energy demand to change the overall trend. However, the LT carbonization showed that this is only true in a limited temperature range as the significant temperature increase in this section from 400–550-750 °C to 450–750-900 °C, necessary to achieve the desired material properties, has a detrimental effect when comparing energy demand on a material output basis. The HT, due to no temperature changes, always shows a clear downward trend for shorter residence times as expected. Throughout the analysis of all heat treatment steps the energy demand of the "Fast" trial was an exemption, driven by the lower overall yield (47%) compared to the other trials (55 ± 2%).

The presented differences in energy demand for varying production throughput are substantial and translate to potentially significant energy and cost savings when the results are virtually scaled to commercial dimensions. Production quantities and energy demands reported by [30] and [29] reflect current typical commercial production scenarios and therefore a combination of the two was used for the following scale-up exercise. Omnia LLC [29] discussed a 2000 ton/year production line with a stable state electric energy demand of approximately 4 MW, while Harper International [30] based their calculations on a 1500 ton/year production facility with an electric energy demand of approximately 6 MW.

For the following calculations and simulation of energy saving a 2000 ton/year production line with an energy demand of 5 MW was therefore used as a basis.

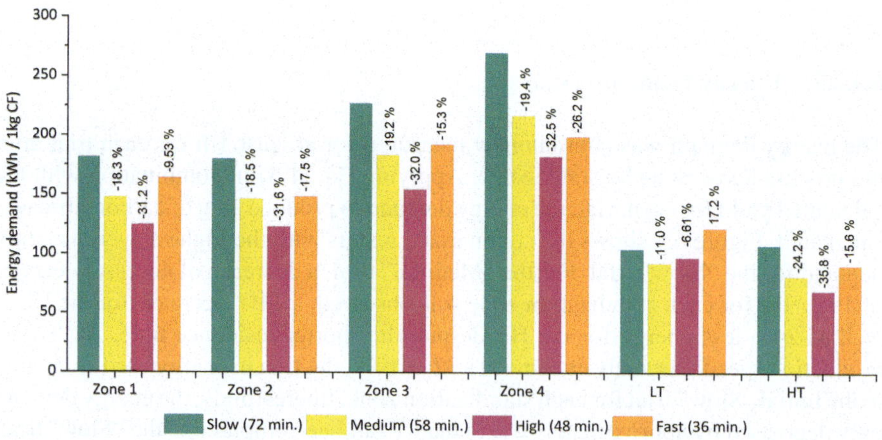

Fig. 6.7 Comparison of energy demand in the different process stages of the four trials per kg carbon fibre. Reproduced from [1] with permission from Elsevier © Elsevier 2023

Furthermore it was assumed that the stabilization accounts for 62% of the overall energy demand, the LT carbonization for 7% and the HT carbonization for a further 10% based on the range reported by literature [29, 30]. The determined energy savings in this research scale study were then extrapolated to industrial scale. From the observations shown in Fig. 6.7, the energy saving potential during stabilization was assumed to be 18% for the "Medium" run across all zones, 32% for the "High" run. The LT carbonization improvement potential was 11% for "Medium", 7% for "High", while HT potential was 24% for "Medium" and 36% for the "High" trial. The "Fast" run was not included in this part of the study as it did not show further energy related improvement in any process section.

At "Medium" production speed, the energy saving was calculated to be between 0.75MW for the 2000 ton/year (5 MW) line and 2.24MW for the 10,000 ton/year (15 MW) scenario, while at "High" production throughput 1.18 – 3.58 MW could be saved for the 2000 ton/year and 10,000 ton/year, respectively, as shown in Fig. 6.8.

The decrease in energy demand during carbon fibre manufacture would significantly reduce greenhouse gas emissions and lead to a cleaner production, with worldwide power generation today still heavily reliant on fossil energy [31]. Based on this study the overall environmental footprint of carbon fibre production and subsequent end products are a function of production speed and therefore fine-tuning the production speed, to a point, while optimising other process parameters such as temperatures and tension to maintain the desired material properties, can significantly improve the materials impact.

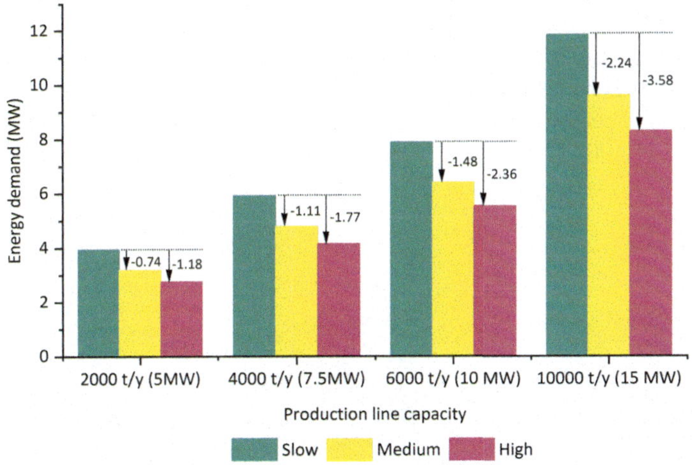

Fig. 6.8 Scaled up overall energy demand and saving potential for the three slower processing rates. Reproduced from [1] with permission from Elsevier © Elsevier 2023

6.4.3.3 Cost Impact Analysis

The cost analysis in this study aimed to highlight the impact of energy saving potential on the overall carbon fibre cost of a 2000 ton/year carbon fibre line. The analysis is based on production costs reported by Nunna et. al [32], where a cost model for carbon fibre production, including precursor manufacturing, was developed. The cost model uses industrial reference prices and knowledge to compile a list of the relevant cost factors and their relative proportion.

Figure 6.9 shows the relative proportions of the contributing cost factors of carbon fibre manufacturing for the three scenarios. The "Slow" scenario displayed the highest fraction of energy costs with 7.1%. Increasing the production speed lowered this portion to 5.8% for the "Medium" setup and further decreased to 5.1% for the "High" scenario. By systematically improving processing conditions, and tailoring these for faster line speeds, a 2% reduction of the share of energy demand was achieved. The increased material throughput is expected to have a cumulative effect on other cost factors, such as labour. Focusing on the impact of improving the process by only adjusting basic processing conditions, this was however not further detailed in this study to highlight the change related to energy costs.

Figure 6.10 depicts the annual carbon fibre production cost and cost savings due to reduction of energy demand with higher production throughput on industrial scale line capacities. The overall annual costs shown here were again based on the cost model of Nunna et. al [32]. While the cost for carbon fibre production increases from 21 to 108 Mio USD annually for line capacities ranging from 2000 to 10,000 ton/year, the relative potential cost savings due to higher production speed seemed relatively small. However, as shown in Fig. 6.10b) the absolute amount of cost saved is significant, considering no additional investment or major changes to the process are necessary to achieve this outcome. For a 2000 ton/year line capacity increasing the manufacturing speed would save approx. 285,000 USD for the "Medium" scenario and approx. 456,000 USD for the "High" scenario. These values increase to 570,000

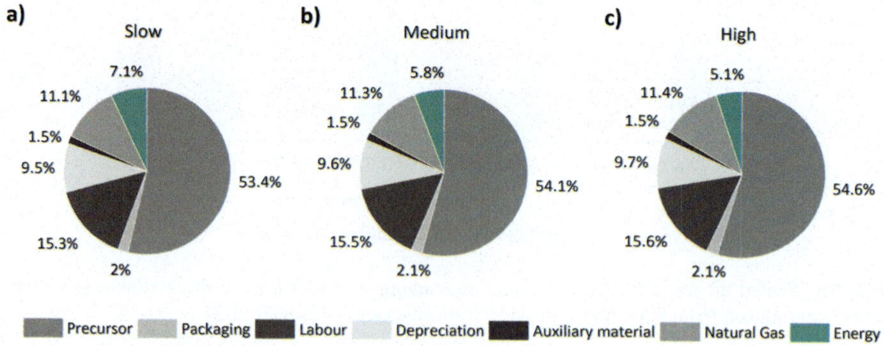

Fig. 6.9 Change of cost proportion of energy demand for the three slower processing conditions based on a reported 2000 ton/year cost model [33]. Reproduced from [1] with permission from Elsevier © Elsevier 2023

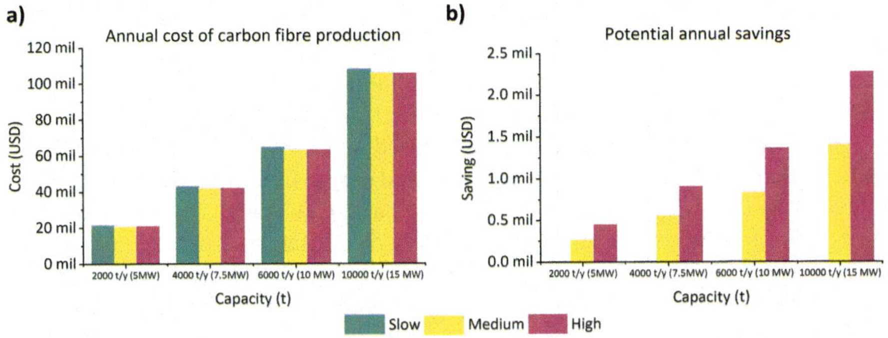

Fig. 6.10 **a** Projected annual cost of CF production for scaled up production capacities and **b** related potential savings per year. Reproduced from [1] with permission from Elsevier © Elsevier 2023

and 913,000 for a 4000 ton/year capacity setup and 856,000 and 1.3 Mio for a 6000 ton/year production line. In a 10,000 ton/year scenario savings add up to approx. 1.4 Mio USD for the "Medium" scenario and 2.3 Mio USD for the "High" scenario, again highlighting the importance of thoroughly optimizing the manufacturing conditions in terms of ideal processing speed and exploring the limits of processing window for a specific precursor.

6.5 Discussion

This chapter focused on the assessment of energy demand and related emissions for the key process steps during the transformation of polyacrylonitrile (PAN) precursor to carbon fibre at different processing speeds while maintaining similar carbon fibre material properties. Each manufacturing step was analysed in detail for varying processing rates and linked to the resulting material properties and generated emissions. Fibre material samples were gathered after the stabilization ovens and carbonization furnaces and corresponding emissions were measured.

Even though the stabilization temperature was increased by approximately 15 °C from the slowest to the fastest production speed, the continuously measured energy demand did not significantly increase. Due to the shorter residence time, the energy used per kg carbon fibre therefore decreased by 19% for the "Medium" trial and 32% for the "High" trial. With a wider temperature spread of at least 50 °C between each of the trials, the energy demand in LT carbonization section decreased by 11% for the "Medium" trial and 6% for the "High" trial, compared to the baseline "Slow" run. When processing the studied precursor fibre at "Fast" speed with the shortest residence time, this type of optimization becomes limited, where the benefits of increased speed are outweighed by other detrimental effects. The main reason for a trend reversal is the significantly lower yield of only 47% compared to an average of 55% for the slower trials, diminishing any potential saving effects attributed to the increased material throughput. Even though temperatures were not changed throughout the study in the HT carbonization furnaces, due to the reduced residence time, energy savings per kg carbon fibre of 24% for the "Medium" trial, 36% for the "High" and 16% for the "Fast" trial were observed.

The emission analysis in this chapter focused on HCN, NH3, CH4 and CO gases and revealed that HCN is the main gas generated during fibre conversion (avg. 194g), followed by NH3 (avg. 63g), CO (avg. 44g) and CH4 (avg. 20g), measured per kg carbon fibre. With less than approximately 5% difference (per kg carbon fibre) between the three slower processing speeds, it was concluded that the process itself did not significantly influence the emission composition in this range. This was further confirmed by the evolution of material properties, which showed very similar changes throughout the process. The "Fast" trial showed, however, that the material yield directly correlates to the amount of emitted gases measured, where the significantly lower material yield correlated to the highest amount of emissions of all four measured species. This was thought to be induced by limited chemical conversion and lower oxygen uptake of the material during stabilization, increasing the mass loss during following rapid carbonization reactions. The high heating rate may have increased polymer defragmentation, chain scission and led to higher emission levels [5, 34].

6.6 Conclusion

This chapter assessed the energy demand and related emissions for the main carbon fibre manufacturing steps when transforming polyacrylonitrile (PAN) precursor to carbon fibre at different processing speeds. For the first time, this work presents a detailed analysis of energy demand for varying production conditions on a continuous production scale, while achieving similar carbon fibre material properties. The insights highlight considerable energy saving potential through exploring the process limits and fine-tuning manufacturing conditions. Lower energy demand directly translates to lower emissions from energy-generation and a more environmentally friendly product. The overarching intention of this study was to generate a better holistic understanding of relationships between processing parameters, resulting energy demand, emissions and the material transformation process.

The study conveys that energy savings due to higher material throughput per hour significantly outweighed the relatively small increase in energy demand due to higher processing temperatures that are required at higher processing speed, which therefore led to an overall decrease in energy demand per kg final product and consequently lower energy-related emissions. The "Fast" trial confirmed processing limits of the investigated precursor material and highlighted the impact of reduced material yield on energy demand per kg final product.

The findings presented in this chapter foster a better understanding of the correlation between processing conditions, material properties, generated emissions and resulting energy demand. These insights enable research to move towards an overall cleaner production, reducing manufacturing cost for both stabilization and carbonization processes and aid to uncover potential improvements to the manufacturing equipment. One such conclusion may be that precursor development towards polymer chemistry for faster processability may yield greater indirect environmental and cost improvements than reducing process temperatures.

Future research in this field could focus on further exploring the practical limits of processability and related implications, safety and limitations of equipment in terms of temperature and emission removal.

References

1. Groetsch T, Maghe M, Creighton C, Varley RJ (2023) Environmental, property and cost impact analysis of carbon fibre at increasing rates of production. J Cleaner Prod 382. https://doi.org/10.1016/j.jclepro.2022.135292
2. Das S (2011) Life cycle assessment of carbon fiber-reinforced polymer composites. Int J Life Cycle Ass 16(3):268–282. https://doi.org/10.1007/s11367-011-0264-z
3. Baker DA, Rials TG (2013) Recent advances in low-cost carbon fiber manufacture from lignin. J Appl Polym Sci 130(2):713–728. https://doi.org/10.1002/app.39273
4. Griffing E (2009) Carbon fiber from PAN. In: Life cycle inventory summary. Wichita State University
5. Chung DDL (1994) Carbon fiber composites (no. 1). Heinemann, Newton, MA, 1994, p 209
6. Dér A et al (2021) Modelling and analysis of the energy intensity in polyacrylonitrile (PAN) precursor and carbon fibre manufacturing. J Clean Prod 303:127105. https://doi.org/10.1016/j.jclepro.2021.127105
7. Park S-J, Heo G-Y (2015) Precursors and Manufacturing of Carbon Fibers. In: Park S-J (ed) Carbon fibers. Springer Series in Materials Science.Springer Netherlands, Dordrecht, Chapter 2, pp 31–66
8. Inagaki M (2000) New Carbons—Control of Structure and Functions. Elsevier, p 229
9. Hiremath N, Young S, Ghossein H, Penumadu D, Vaidya U, Theodore M (2020) Low cost textile-grade carbon-fiber epoxy composites for automotive and wind energy applications. Compos Part B: Eng 198. https://doi.org/10.1016/j.compositesb.2020.108156
10. Ennis BL et al (2019) Optimized carbon fiber composites in wind turbine blade design. Sandia National Lab.(SNL-NM), Albuquerque, NM (United States)
11. Nguyen-Thai NU, Hong SC (2013) Structural Evolution of Poly(acrylonitrile-co-itaconic acid) during Thermal Oxidative Stabilization for Carbon Materials. Macromolecules 46(15):5882–5889. https://doi.org/10.1021/ma401003g
12. Nunna S et al The effect of thermally induced chemical transformations on the structure and properties of carbon fibre precursors. J Mater
13. Nunna S et al (2019) Time dependent structure and property evolution in fibres during continuous carbon fibre manufacturing. Materials (Basel) 12(7):1069. https://doi.org/10.3390/ma12071069
14. Bajaj P, Roopanwal AK (1997) Thermal stabilization of acrylic precursors for the production of carbon fibers: an overview. J Macromolecular Science, Part C: Polymer Rev 37(1):97–147. https://doi.org/10.1080/15321799708014734
15. Ge Y, Fu Z, Zhang M, Zhang H (2020) The role of structural evolution of polyacrylonitrile fibers during thermal oxidative stabilization on mechanical properties. J Appl Polym Sci 138(1):49603. https://doi.org/10.1002/app.49603
16. Takaku A, Hashimoto T, Miyoshi T (1985) Tensile properties of carbon fibers from acrylic fibers stabilized under isothermal conditions. J Appl Polym Sci 30(4):1565–1571. https://doi.org/10.1002/app.1985.070300421

17. Hanna SB, Yehia AA, Ismail MN, Khalaf AI (2012) Preparation and characterization of carbon fibers from polyacrylonitrile precursors. J Appl Polym Sci 123(4):2074–2083. https://doi.org/10.1002/app.34704
18. Ji M, Wang C, Bai Y, Yu M, Wang Y (2007) Structural evolution of polyacrylonitrile precursor fibers during preoxidation and carbonization. Polym Bull 59(4):527–536
19. Ko T-H (1991) The influence of pyrolysis on physical properties and microstructure of modified PAN fibers during carbonization. J Appl Polym Sci 43(3):589–600. https://doi.org/10.1002/app.1991.070430321
20. Ko T-H, Huang L-C (1998) The influence of cobaltous chloride modification on physical properties and microstructure of modified PAN fiber during carbonization. J Appl Polym Sci 70(12):2409–2415. https://doi.org/10.1002/(sici)1097-4628(19981219)70:12%3c2409::Aid-app13%3e3.0.Co;2-0
21. Morgan P (2005) Carbon fibers and their composites. CRC press, Ch 1174, pp 185–259
22. Kaur J, Millington K, Smith S (2016) Producing high-quality precursor polymer and fibers to achieve theoretical strength in carbon fibers: A review. J Appl Polym Sci 133(38):43963. https://doi.org/10.1002/app.43963
23. Hameed N et al (2016) Structural transformation of polyacrylonitrile fibers during stabilization and low temperature carbonization. Polym Degrad Stab 128:39–45. https://doi.org/10.1016/j.polymdegradstab.2016.02.029
24. Goodhew PJ, Clarke AJ, Bailey JE (1975) A review of the fabrication and properties of carbon fibres. Mater Sci Eng 17(1):3–30. https://doi.org/10.1016/0025-5416(75)90026-9
25. Donnet J-B, Chand Bansal R (1998) Carbon fibers. CRC Press, ch. 572, pp 1–370
26. Rahaman MSA, Ismail AF, Mustafa A (2007) A review of heat treatment on polyacrylonitrile fiber. Polym Degrad Stab 92(8):1421–1432. https://doi.org/10.1016/j.polymdegradstab.2007.03.023
27. Zhu D, Xu C, Nakura N, Matsuo M (2002) Study of carbon films from PAN/VGCF composites by gelation/crystallization from solution. Carbon 40(3):363–373. https://doi.org/10.1016/s0008-6223(01)00116-6
28. Laffont L, Monthioux M, Serin V, Mathur RB, Guimon C, Guimon MF (2004) An EELS study of the structural and chemical transformation of PAN polymer to solid carbon. Carbon 42(12–13):2485–2494. https://doi.org/10.1016/j.carbon.2004.04.043
29. LLC O (2012) Assessment of carbon fiber manufacturing cost—extract from a benchmarking study on Iceland as a location for carbon fiber production. Omnia LLC, Iceland
30. Fry J (2016) The future of high capacity carbon fiber production. In: Paris, H, International (eds) JEC world, 2016: Harper International, pp 1–37. [Online]. Available: http://www.harperintl.com/wp-content/uploads/2011/02/The-Future-of-High-Capacity-Carbon-Fiber-Production-Lines-Harper-International.pdf. [Online]. Available: http://www.harperintl.com/wp-content/uploads/2011/02/The-Future-of-High-Capacity-Carbon-Fiber-Production-Lines-Harper-International.pdf
31. Kober T, Schiffer HW, Densing M, Panos E (2019) Global energy perspectives to 2060 – WEC's World Energy Scenarios 2019. Energy Strategy Reviews, vol 31. https://doi.org/10.1016/j.esr.2020.100523
32. Nunna S, Blanchard P, Buckmaster D, Davis S, Naebe M (2019) Development of a cost model for the production of carbon fibres. Heliyon 5(10):e02698. https://doi.org/10.1016/j.heliyon.2019.e02698
33. König S et al (2021) High-performance carbon fibers prepared by continuous stabilization and carbonization of electron beam-irradiated textile grade polyacrylonitrile fibers. Macromolecular Mater Eng 306(12). https://doi.org/10.1002/mame.202100484
34. Grassie N, McGuchan R (1971) Pyrolysis of polyacrylonitrile and related polymers—III. Thermal analysis of preheated polymers. Eur Polymer J 7(10):1357–1371. https://doi.org/10.1016/0014-3057(71)90032-2

Chapter 7
Variation of Precursor Material

The chemistry and structure–property relationships of precursor materials for carbon fibre manufacture have been and still are extensively studied. However, no detailed studies are available in literature that compare different types of precursors on a continuous process scale and how they may affect energy demand, environmental impact and cost. An in-depth understanding of the effect the precursor material has on the selection of process parameters and resulting carbon fibre properties, energy demand, environmental impact and how these factors influence each other can provide new insights into process optimization and cost saving potentials for carbon fibre. Furthermore, these findings can assist outlining key areas to focus on when assessing the potential of future precursor alternatives. The primary goal of this study was to gain a better understanding of how the choice of precursor material affects processing conditions and to highlight links between the material transformation, energy usage, emission composition and the rate of manufacture. For the first time this work provides a comparison of three precursor materials in terms of cost, environmental impact, processing and material performance on a large continuous manufacturing scale.

7.1 Background

PAN fibres are currently the predominant precursor type due to the high yield that can be achieved along with the possible high-performance properties of the final product [2, 3]. Polyacrylonitrile has a carbon content of 67.9%, which makes it an ideal material to carbonize with an overall achievable yield of typically above 50% [2, 4]. The Special Acrylic fibre (SAF) precursor in use today was developed from material originally intended for textile products [2]. The high cost and long

This chapter (modified) was published in [1], Copyright Elsevier 2023.

T. Groetsch, *Enhancing Carbon Fibre Production Through Life Cycle Analysis*, Springer Theses, https://doi.org/10.1007/978-3-031-70578-6_7

necessary processing times, however, make PAN a relatively expensive precursor material for carbon fibres. PAN degrades well before its melting point and therefore cannot easily be melt spun [5]. A more complex and expensive solution-spinning technique is commonly used to produce fibre material [2, 6]. PAN homopolymer shows a fast evolution of heat when exothermic reactions start to occur, which makes controlling the reaction difficult [2, 7, 8]. Adding acidic comonomers, such as itaconic acid, can lower the initial reaction temperature, improve the exotherm and increase carbon yield [9, 10]. In order to enhance processability in the spinning stage and improve the material's morphology, neutral comonomers, such as methyl acrylate, are added [9]. Commercially available precursor fibres typically contain a combination of comonomers for improved fibre performance, typically below 10%, as higher contents would decrease the material's properties or inhibit processability [2, 6, 7, 11, 12]. The thermal stabilization of the PAN fibres is carried out at 200–300 °C at slow heating rates, leading to a very time and energy intensive process [13, 14]. Various studies have been undertaken detailing new or improved approaches to decrease the necessary oxidation time and increase the output [15–23].

All approaches aimed to reduce the overall energy required and subsequent cost and environmental impact for the final product. In most cases, however, additional research is necessary in order to prove and implement these ideas on an industrial scale. Furthermore, with a significant contribution of the reported SAF precursor (over 50%) to the overall cost of carbon fibre, the potential to reduce cost and environmental impact of the final product is limited [24, 25].

Due to the high price and difficult manufacture of SAF PAN typically used for carbon fibre production, other acrylic fibres, commonly used for textile products, may be viable alternatives to reduce the overall cost of the final product [11]. In the context of this study, textile PAN refers to polyacrylonitrile-based precursor with a different chemical composition than traditionally used SAF PAN precursor, that promises improvements in terms of cost, energy and/or environmental impact. Textile grade PAN material generally contains at least 85% acrylonitrile, the remainder of the material is made up of comonomers, up to 15% [11, 26]. The added comonomers serve to improve processability and tune the material to the desired application. The comonomer ratio in textile fibres is typically determined for economic reasons and specific target characteristics such as dye uptake or better solubility in spinning solvents [11, 27]. Comonomers such as methyl acrylate and methyl methacrylate are added to improve the solubility in solvents and overall spinnability of PAN [9, 10]. Acidic comonomers, such as itaconic acid, can be added to improve dye uptake and to increase hydrophilicity and improve densification [5, 10, 28]. When carbonizing the material, however, higher amounts of comonomers and impurities lead to increased mass loss and negatively affect the achievable mechanical properties [29]. There has been substantial research effort over the years investigating the viability of textile precursor for carbon fibre manufacture [11, 27, 29–35]. The reviewed approaches intended to reduce cost for the final product and even though some of these concepts incorporated additional (pre-)treatment steps or alternative processing equipment, all concluded with varying levels of cost reductions. All ideas however still need to be proven economically and environmentally feasible on an industrial scale.

Increased environmental awareness and considerations have drawn focus on more sustainable alternatives based on plant material, such as lignin and/or cellulose [36, 37]. These bio-based materials are more sustainable alternatives to the commonly used PAN precursor material, with potential environmental as well as economic benefits. The use of lignin, a macromolecule present in about a third of woody plants, may have the potential to save almost 50% of the cost of carbon fibres [38]. As lignin is a by-product in processes such as paper production, there is an abundant supply available at low cost. The high molecular carbon content makes it an ideal precursor candidate. There are, however, some limitations, that to date have prevented a more widespread use of the material in carbon fibre production [39]. Bajpai et al. [38] published a review on processability and material properties of lignin from different plant sources and extraction routes (e.g. Kraft process, Steam Explosion, Organosolv). While the reviewed publications showed many different processing and improvement approaches and results vary, there is common ground when comparing to PAN-based carbon fibres. The tensile strength and modulus of lignin-based carbon fibres (<1 GPa, > 80 GPa) is much lower than that of PAN-based fibres (>3GPa, < 100 GPa), while the average diameter is higher (Lignin > 10μm, PAN < 10μm) [38]. Due to the large variety of feedstock, the properties of lignin-based carbon fibres can greatly vary, while for PAN-based carbon fibres the values are in a much smaller range [40–42]. As for precursor fibre production, lignin is more versatile than PAN, as it can be wet spun, dry spun and melt spun [43, 44]. With lignin being a natural material obtainable from a variety of sources, purity is another key factor for processability of the material [43]. With improvement of the source material in terms of purity and material properties, and some processing adjustments, lignin shows great potential as a precursor for carbon fibre production [11, 38, 45–47].

Another route of improving the shortcomings of the material is to blend it with other materials. Cellulose was investigated in various studies, as the material also is fully renewable [36, 37, 40, 48–50]. While cellulose-based precursor fibre can be converted into high quality carbon fibre material with mechanical properties as high as 3.5 GPa tensile strengths and an initial tensile modulus of up to 690 GPa using tension and very high temperatures [2], the main limitation of cellulose, however, is its low carbon fibre yield of around 10–30% owing to relatively low initial carbon content of 44% [36]. Lignin on the other hand has a significantly higher carbon content of over 60%, but due to the complex and variable molecular structure, achievable mechanical properties are lower than for cellulose [37]. Various researchers investigated blending lignin and cellulose to improve the material's suitability for carbon fibre production, studying the effect of mixing ratio and processing on material properties [36, 37, 40, 42, 48, 50]. Even though lignin/cellulose-based carbon fibres have not reached the mechanical properties of PAN-based fibres yet, the potential of lignin/cellulose blends as a precursor material remains considerable. The low cost, bio-based and fully renewable material is considered a great alternative for applications with higher monetary, less mechanical property focused requirements.

7.2 Materials and Methods

Even though different sources of materials were used, the overall goal was to produce carbon fibres from the three materials with similar mechanical properties (≈3.0 GPa tensile strength, ≈ 210 GPa initial tensile modulus, mass density 1.7–1.8 g/cm3.). This, however, was only achievable for the first two materials, special carbon fibre grade PAN and textile PAN. The cellulose-lignin-blend, due to being still in an early experimental phase, did not reach mechanical properties exhibited by the other two precursors. This has to be kept in mind for any comparisons made in this study.

7.2.1 Material Properties

Three carbon fibre precursor materials were investigated, special acrylic fibre (SAF PAN), textile PAN (TexPAN) fibre and lignin/cellulose blend (LigniCel). The SAF PAN used in this study is specifically produced for carbon fibre manufacture with the chemical composition comprising Acrylonitrile (AN), methyl acrylate (MA) and itaconic acid (IA) to improve processability during the stabilization stage. The comonomer composition of the textile PAN (TexPAN) fibre was not tailored to improve processability in the same way. Both these materials had a tow size of 12,000 filaments (12k) unlike earlier trials outlined in previous chapters, which may introduce small uncertainties when comparing results between chapters.

The third material investigated was an experimental dry-jet wet spun hardwood lignin/cellulose blend (LigniCel) with a ratio of 50:50. The cellulose pulp had a molecular weight of 163.6 kg/mol and the Lignin a molecular weight of 2.23 kg/mol. The material was stringently filtered to produce the desired material properties. As a bio-derived material available from a variety of sources, the quality and purity of the initial source material may have a significant effect on the achievable material properties. Variations thereof are however not considered in this work but should be kept in mind when assessing these results. Due to the experimental nature of the material the largest tow size available was only 400 filaments (0.4 k). Results shown here were therefore generated on benchtop equipment, since continuous processing was not feasible for a comparison study due to the small tow size. Comparison to the other materials was enabled by theoretically assessing the process, energy demand and cost based on literature and up-scaled values.

In order to have a basis for evaluation, carbon fibres with comparable mechanical properties were assessed, dictated by the maximum achievable properties for the TexPAN material with an optimized setup. To produce a comparable material grade, the PAN process was optimized towards lower energy demand while maintaining similar properties to what was achieved for the TexPAN. The targeted values for the most relevant material properties were a tensile strength of 3.0 ± 0.2 GPa, an initial tensile modulus of 210 ± 6 GPa and a mass density in the range of 1.7–1.8 g/cm3. Due to the smaller tow size, LigniCel was processed at smaller scale than the other

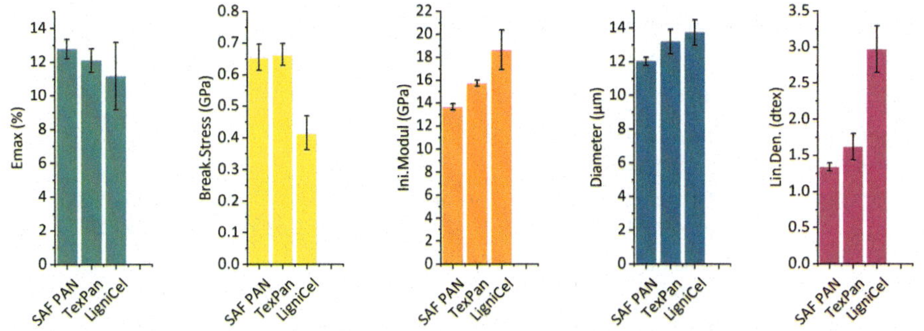

Fig. 7.1 Comparison of material properties for the three analysed precursors, PAN, TexPAN and LigniCel. Reproduced from [1] with permission from Elsevier © Elsevier 2023

two precursors, nevertheless the benchtop scale results generated can serve as a first indication of viability of the material.

Figure 7.1 shows selected material properties for the three analysed precursors. LigniCel exhibits the lowest maximum elongation and strength of 11% and 0.42 GPa, respectively. Initial tensile modulus (18.65 GPa), linear density and diameter, however, are all larger than observed for both PAN-based fibre types. These relatively low carbon fibre properties, however, show that the conversion process is not yet at the same level of industrialization as it is for PAN-based material.

7.2.2 Carbon Fibre Manufacturing Process

The process data used in this chapter was collected from the research manufacturing line at Carbon Nexus, located at Deakin University (Australia), in the same way as outlined in previous chapters [16, 51, 52].

7.2.3 Processing Conditions

Table 7.1 shows the applied processing conditions for the analysed precursor materials, optimised to achieve the desired mechanical properties mentioned above. As LigniCel was not available for extensive experimental evaluation on the required scale, the assessment uses a combination of small-scale testing and a theoretical approach based on literature estimations.

Table 7.1 Optimized processing conditions for each analysed alternative precursor

Stage	Parameters	Values		
	Reference name	PAN	TexPAN	LigniCel[a]
Stabilization (oxidation)	Total residence time in heat (min)	58	96	61
	Temperature profile per oven (°C)	234–243 to 253–262	235–247 to 254–263	245–255 to 265–273
	Tension profile per oven (N)	18–18 to 20–19	14–14 to 15–12	–
Low temp. carbonization (LT)	Total residence time in heat (min)	3.6	6	10.25
	Temperature profile per zone (°C)	400–550 to 700	400–600 to 800	400–505 to 630
	Tension (N)	11	18	–
High temp. carbonization (HT)	Total residence time in heat (min)	2.4	4	19.75
	Temperature profile per zone (°C)	1000–1300	1100–1400	1000–1250
	Tension (N)	18	21	–

[a]Theoretical estimations based on [40, 42] using similar source material

7.2.4 Assessment Approach

The same streamlined assessment approach was used as described in previous chapters. Figure 7.2 shows the manufacturing process including the breakdown into main processing steps for the PAN-based and modelled LigniCel production processes.

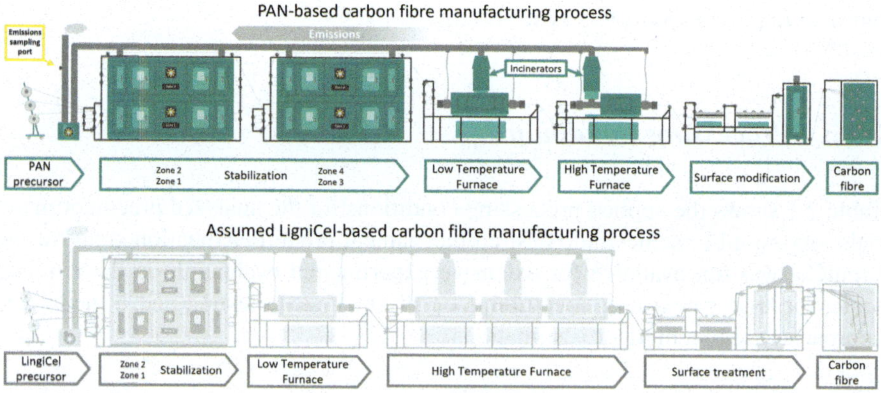

Fig. 7.2 Carbon fibre manufacturing process for different materials with main production stages. Reproduced from [1] with permission from Elsevier © Elsevier 2023

The presented research again focuses on a number of key areas to compare the different precursor scenarios: material properties as well as energy demand, emissions and cost per kilogram carbon fibre. The parameters referred to for the final output material properties were, in line with previous chapters, the commonly referenced tensile strength, initial tensile modulus and overall yield.

7.2.5 Energy Measurement and Material Characterisation

Energy demand was recorded and calculated according to the procedures outlined earlier. The material was characterized in the same way as in previous chapters. Additionally thermogravimetric analysis (TGA) was performed to analyse the materials.

7.3 Results and Discussion

The precursor material used to produce carbon fibres has great influence on the properties of the produced fibres, production parameters, yield and the overall cost of the final product [11]. The most commonly used precursor in industrial scale production today is SAF PAN [5, 24]. To improve the cost structure of carbon fibre, textile grade, low-cost PAN is being investigated and developed as a potential replacement for the more expensive, specifically modified PAN used in current commercial production [30, 34, 53]. With sustainability in mind and the aim to further lower cost, research is also focusing on bio-based material, of which lignin and cellulose currently are the most common ones [37, 40, 42, 49, 50]. Various other types of precursor materials, e.g. pitch and polyethylene, have been developed or are already in use, but either their production volume, achievable material properties, cost or the development phases precludes these materials from being an alternative in the above-mentioned context [54, 55].

7.3.1 Precursor-Specific Production Processing Scenarios

The temperature–time profiles shown in Fig. 7.3 were derived by optimizing the process to reach the desired material properties ($\approx$3.0 GPa tensile strength, $\approx$ 210 GPa initial tensile modulus), for both the SAF PAN and TexPAN fibres. Temperatures ranged from 234 °C to 262 °C during stabilization for the SAF PAN reference material, with a residence time of 58 min. The temperature-profile for TexPAN was very similar to the reference material ranging from 235 °C to 263 °C, however, the time spent in heat during stabilisation increased by 66% to 96 min to help control exothermic reactions, mainly the conversion from nitriles into heterocyclic structures

[2]. The processing conditions for LigniCel derived from [40, 42] are presented for comparison. The theoretical estimations for LigniCel based on average values from recent literature [40, 42] suggested a stabilisation temperature profile of 245–272 °C with only 61 min residence time. The carbonization temperature profile for SAF PAN and TexPAN was again similar with 1000–1300 °C for PAN and 1100–1400 °C for TexPAN. Due to the continuous nature of the process and utilizing the same processing equipment, the time in heat during carbonization was again 66% longer for TexPAN. Research [40, 50] suggests that in order to produce relevant material grade carbon fibres from LigniCel, longer carbonization times may be necessary. The theoretical process specifications suggested here therefore assumed longer carbonization times than would be achievable with the equipment used for the other materials in a continuous process. The total carbonization time was proposed to be 30 min for LigniCel opposed to 6 min for the SAF PAN reference material. The temperature range of the LigniCel process is slightly different to the other two materials with a range of 400–1600 °C. The higher carbonization temperatures are required to improve the final carbon fibre mechanical properties. It has to be emphasized that this is just a theoretical estimation based on recent literature and the process does not produce the same mechanical quality of material achievable with SAF PAN and TexPAN.

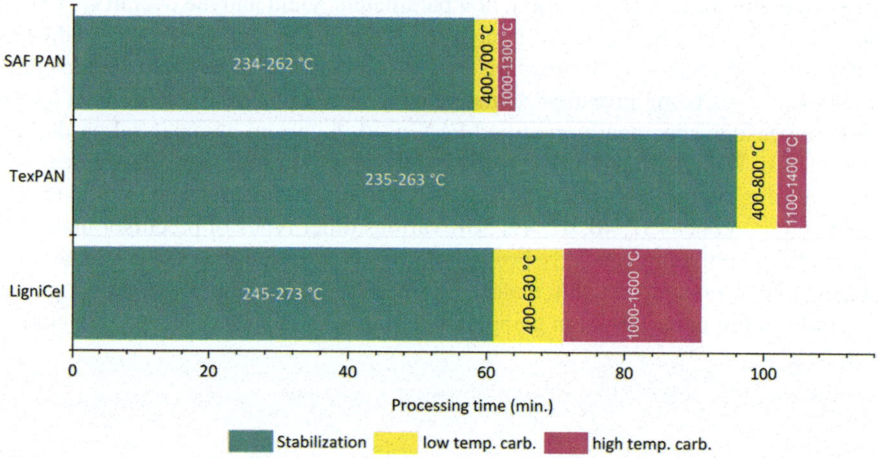

Fig. 7.3 Variations in time–temperature profiles for three different precursor materials during stabilization and carbonization. Reproduced from [1] with permission from Elsevier © Elsevier 2023

7.3.2 *Mechanical Properties*

Figure 7.4 depicts a comparison of mechanical properties of SAF PAN and TexPAN fibres along the manufacturing process. The tensile strength was very similar for both precursors (0.66 GPa), while the initial modulus was found to be 13.68 GPa and 15.74 GPa for SAF PAN and TexPAN, respectively. Both precursors showed a drop in strength and modulus after stabilization, testimony of significant chemical and structural changes. After stabilisation into OPF, the tensile strength dropped to 0.35 GPa for SAF PAN and 0.25 GPa for TexPAN fibres. Similarly, the initial modulus of PAN was 16% higher compared to TexPAN (9.95 vs. 8.40 GPa). The removal of non-carbonaceous elements and continued reorganization of the microstructure into more crosslinked graphitic layers and subsequent turbostratic structure led to a tensile strength of 3.13 and 2.93 GPa and an initial tensile modulus of 215.30 and 205.66 GPa for PAN and TexPAN respectively, showcasing the commercial relevance of the produced material [2, 56]. In comparison LigniCel-based carbon fibres were reported to reach values of around 1 GPa ultimate tensile strength and 76 GPa initial tensile modulus, far lower than PAN-based fibres [40, 50, 57].

7.4 Extent of Cyclization Reaction and Fibre Mass Density

For SAF PAN and TexPAN, samples were collected and analysed after each manufacturing step. Figure 7.5a) depicts the extent of cyclization reaction (EOR) for the four stabilization zones outlining the progress of one of the key reactions occurring during the stabilization process. The amount of cyclised material steadily increased

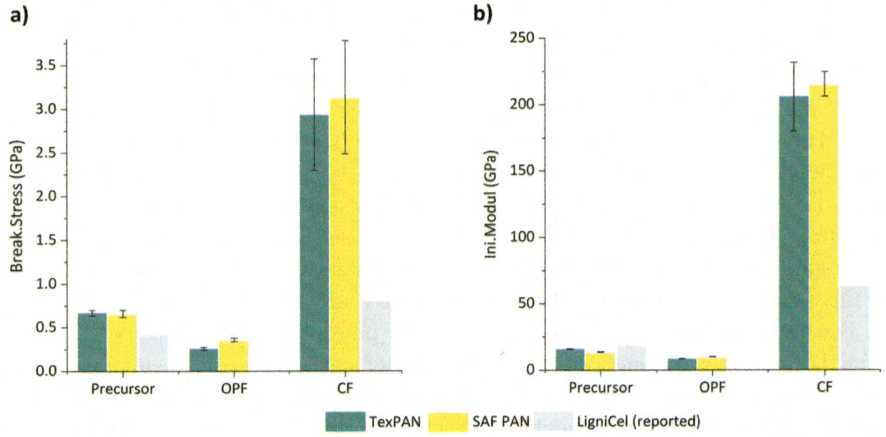

Fig. 7.4 Comparison of mechanical properties for precursor, stabilized fibres and final carbon fibres. Reproduced from [1] with permission from Elsevier © Elsevier 2023

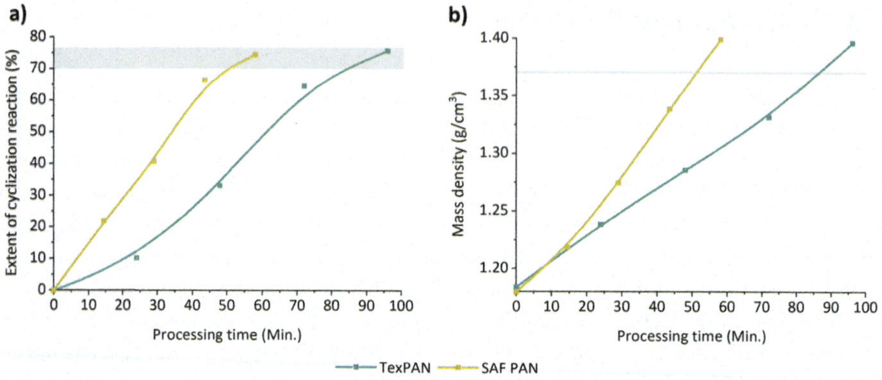

Fig. 7.5 A) Extent of cyclization reaction and b) mass density throughout stabilization for PAN and TexPAN. Reproduced from [1] with permission from Elsevier © Elsevier 2023

throughout zones 1–3 and slowed down slightly in zone 4, which is considered to be a result of less unreacted structures available to participate in the ring-closure reaction [41, 58]. The EOR for the reference PAN material reached 74.5% after the final stabilization stage, while the TexPAN material converted a slightly higher amount of 75.8%. An EOR in this range was targeted, as it is known to correlate with the desired density range after stabilization and a fibre material well prepared for carbonization [41, 58]. The density of the fibre material increased concurrently throughout the stabilization process with both fibres reaching 1.39 ± 0.01 g/cm^3, in the range of previously reported values [41, 51, 59]. The simultaneous increase of the EOR and densification of both fibre materials was thought to be a result of oxygen uptake during stabilization as well as a tight repacking of the fibre's microstructure [60, 61].

Due to the different chemical structure of a lignin-cellulose blend, the concept of increasing extent of cyclization reaction is not a suitable method to determine sufficient stabilization progress [39, 62]. Literature suggests that from peaks at 1740 cm^{-1} and 1430 cm^{-1} however the progress of stabilization can similarly be traced [63]. The increase of mass density could, however, still be used as a metric to establish the progress of the treatment. Lignin typically has a density in the range of 1.35–1.50 g/cm^3 [64] while cellulose is typically measured at 1.50 g/cm^3 [65], both significantly higher than what is typical of PAN-based material (1.18 g/cm^3), even after stabilization. The 50/50 blend precursor material used for experimental work in this study was measured to have a density of 1.42 g/cm^3.

For both PAN-based materials, the mass per unit length/linear density was determined by cutting one-meter-long samples after each stage of the manufacturing process and carefully weighing them. As seen in previous studies [41, 58] the linear density of PAN drops in Zone 1 due to spin-finish and moisture removal as well as stretching of the material (Fig. 7.6). The material then exhibits a steady increase in linear density throughout stabilization due to microstructural reorganization and oxygen intake before a significant drop during carbonization due to the removal

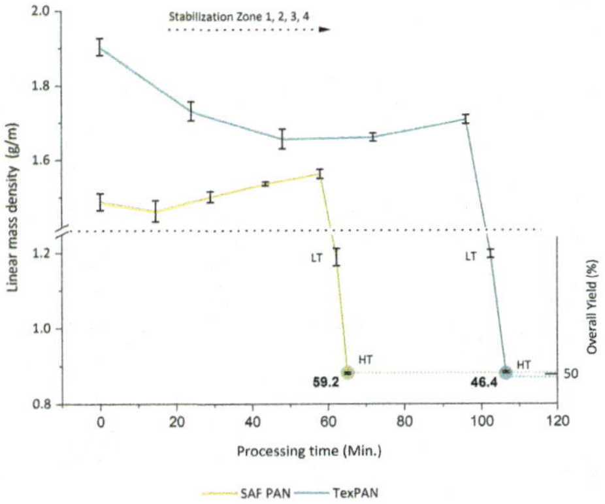

Fig. 7.6 Evolution of linear mass density throughout conversion process and final yield correlated to the final density of the carbon fibre for PAN and TexPAN. Reproduced from [1] with permission from Elsevier © Elsevier 2023

of almost all non-carbon material [58]. The treatment at high temperature leads to significant loss of material through emissions and concludes with an overall carbon fibre yield of 59.2% based on the initial linear density of SAF PAN precursor. While the overall trend was similar for TexPAN, significant differences were observed during stabilization, where the linear density decreases more than halfway through the stabilization process, a potential sign of missing acidic comonomers present in SAF PAN to facilitate a more rapid and controlled stabilization, i.e. cyclization process. Furthermore, the textile fibre does not show the same level of longitudinal shrinkage, at lower processing temperatures, as the special grade material, which may also add to a more extended drop [66]. Throughout carbonization, the material also shows a significant reduction in linear density leading to an overall yield of 46.4%. The lower yield compared to SAF PAN again highlighted the importance of the initial chemical composition, where specific comonomer types improve initial processability, which in turn leads to less material being removed during carbonization. Another cause may be that even the significantly lower processing speed was already causing too aggressive treatment of the fibres and hence the yield dropped [41].

The linear density of the LigniCel material was measured to be 5.71 g/m (based on a 0.4 k tow and theoretically upscaled to 12 k). The higher linear density of this material is caused by a combination of higher mass density and greater diameter of the individual filaments. No intermediate measurements were taken for this material as it could not be processed the same way as the other two. However, based on a yield of 29% reported in literature [40] for a similar 50/50 LigniCel blend, the final carbon fibre was estimated to have a linear density of 1.66 g/m for 12 k tow size, almost twice that of PAN-based carbon fibre of the same tow size.

In order to verify literature values and better estimate what level of yield can be expected when processed on a continuous scale, thermogravimetric analysis (TGA) was performed for all three materials. The analysis was performed simulating the time–temperature profile the fibre would undergo during carbon fibre production, in air atmosphere during the stabilization profile and nitrogen atmosphere during carbonization. The two PAN based materials show a good correlation with actual

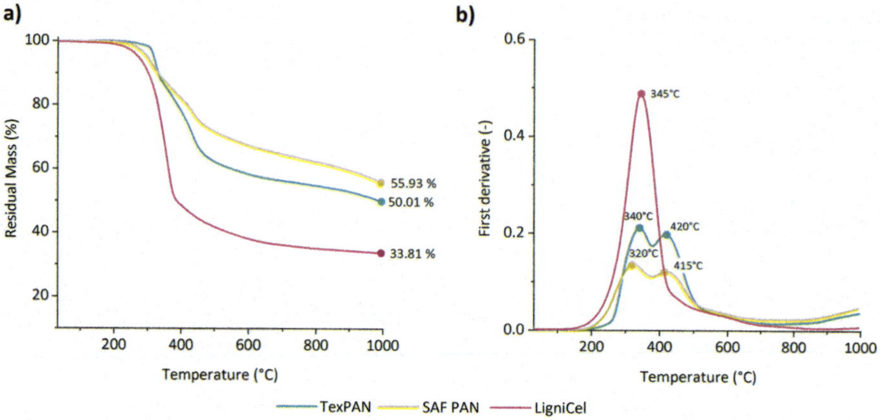

Fig. 7.7 A) Comparison of TGA results for three materials and b) first derivate of results. Reproduced from [1] with permission from Elsevier © Elsevier 2023

results measured at larger scale and shown in Fig. 7.7. With less than 5% difference between TGA results at 1000 °C and the final yield calculated for the large-scale trials, this testing technique can serve as a sufficient yield estimate for the LigniCel material. While the yield found for this material is significantly lower than the PAN-based fibres at only approximately 34%, the results are in line with values reported in the literature [40, 50, 67]. The first derivate curves for SAF PAN and TexPAN show strong similarities with two clearly visible peaks representing the two-step conversion process reported elsewhere [16, 68]. Visible in Fig. 7.7b, the TexPAN peaks occur at slightly higher temperatures, while also being more pronounced, which can be linked to the slower necessary processing due to more retarded reaction mechanisms mentioned earlier. For Lignin however, only one peak is visible, which indicates different conversion mechanisms and in turn a potentially different manufacturing process as outlined earlier.

7.4.1 Elemental Composition of Fibres

Elemental analysis was carried out on fibres collected for each manufacturing stage for SAF PAN and TexPAN (Fig. 7.8) in order to investigate differences in the evolution of chemical structure for the studied materials. The analysis of the SAF PAN fibre revealed an elemental composition consisting of C (66%), N (25%), and H (5.8%) and other elements (2.8%) believed to be mainly oxygen, consistent with findings reported elsewhere [68–71]. During stabilization the C, H, and N content slightly decreased, which was thought to be caused by the formation of gaseous by-products incurred during stabilization [69, 72]. The initial O-content of 2.8%

increased to approximately 15% by the end of stabilization, complementing findings reported elsewhere [73]. Oxygen is the key trigger for the dehydrogenation and oxidation reactions primarily thought to enhance the thermal stability of the fibre during subsequent carbonization [74]. The increasing oxygen content directly correlated with an increasing cyclized structure as shown in Fig. 7.5a. TexPAN started with a composition of C (66%), N (24%), and H (5.9%), evidence of a very similar initial composition of the material in comparison with the SAF PAN. Throughout the conversion process and after carbonization, both precursors show only little difference in terms of their overall elemental composition, with the same amount of C present in the final carbon fibre structure (93%). Interestingly though the O content showed an initial decrease (-2.2%) for TexPAN during stabilization unlike PAN which increased by 0.8% during the same period. This was thought to be caused by the removal of non-reactive comonomers during the initial stages of stabilization.

Figure 7.8b depicts a direct comparison of the elemental composition of the three precursor materials. While the two PAN-based materials are expectedly very similar, LigniCel has a significantly different composition, owing to its different origin and microstructure. Almost no N is present in LigniCel, while oxygen and other elements, such as sulphur and nitrogen, make up 44% of the material, in line with findings reported elsewhere [75]. Also the amount of carbon (50%) is significantly lower than in the other materials, which can explain the lower reported yield achievable with this material [36].

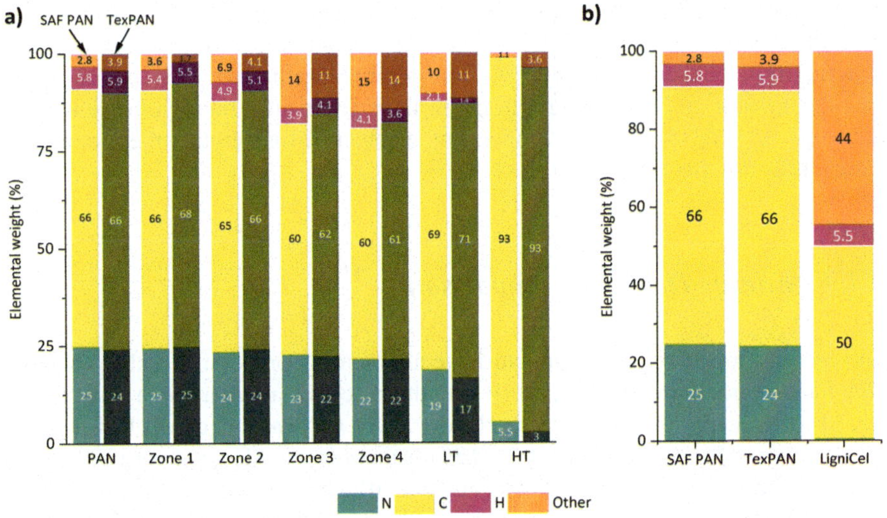

Fig. 7.8 Elemental analysis results for SAF PAN, TexPAN and LigniCel. Reproduced from [1] with permission from Elsevier © Elsevier 2023

7.4.2 Gaseous Emissions

Gases emitted during the conversion from SAF PAN and TexPAN precursor to carbon fibre were collected and analysed to examine differences in amounts and composition. Figure 7.9 shows that for both materials the four most significant gases are hydrogen cyanide (HCN), ammonia (NH_3), carbon monoxide (CO) and methane (CH_4), displayed as a normalized amount of gasses emitted during the production of 1 kg of carbon fibre. Although considered a significant part of the emission's composition, carbon dioxide (CO_2) was not included in the analysis as the manufacturing process is an open system with gases dynamically entering and exiting the equipment. Therefore gases present in standard atmosphere, such as CO_2, cannot be accurately attributed to the process. Even though both precursors are PAN-based, significant differences in the amount of emitted gases were observed, with TexPAN having substantially more gas emitted per kg carbon fibre produced, which corresponds to the lower yield of 46.4% (Fig. 7.9). Out of the studied gases, HCN was the main gas emitted for both materials, with TexPAN emitting approximately 2.9 times more HCN than during the conversion from PAN to carbon fibre. The differences for NH_3, CO and CH_4 were slightly less pronounced, but still were around 2.6 times more than SAF PAN. Notably, while for PAN the second largest amount emitted is CO, for TexPAN it is NH_3. Methane is again the smallest contributor of the four gases for both materials. Due to a less oriented molecular backbone, the commonly referenced conversion mechanism, cyclization of the nitrile groups and cross-linking of the chain-molecules, do not progress as easily and quickly as for the SAF material. Furthermore, the typically higher comonomer content reduces the overall available material to be converted [2, 58, 76]. This in turn, together with longer processing times and lower yield, is thought to cause the significantly higher emission levels observed for TexPAN. Potential underlying mechanisms generating these emissions were outlined in Sect. 5.3.2.

7.4.3 Energy Demand Comparison

The energy demand of the manufacturing equipment for processing both PAN-based materials was measured for at least 1h after the process reached a stable state. Subsequently, the energy necessary to produce 1 kg of CF was calculated based on the yield shown in Fig. 7.6 and Eq. (1.3). The resultant energy demand per process step depicted in Fig. 7.10 and shows the significant difference for the two precursor materials, SAF PAN and TexPAN. The remarkably longer processing time for TexPAN (96 min) compared to SAF PAN (58 min), combined with a lower overall yield significantly impact the energy demand during every stage of the process. Using TexPAN as a source material, and depending on the heat treatment step, 62–70% more energy is required to produce 1 kg of CF compared to SAF PAN. A small contribution to the overall higher energy demand may also be due to the slightly

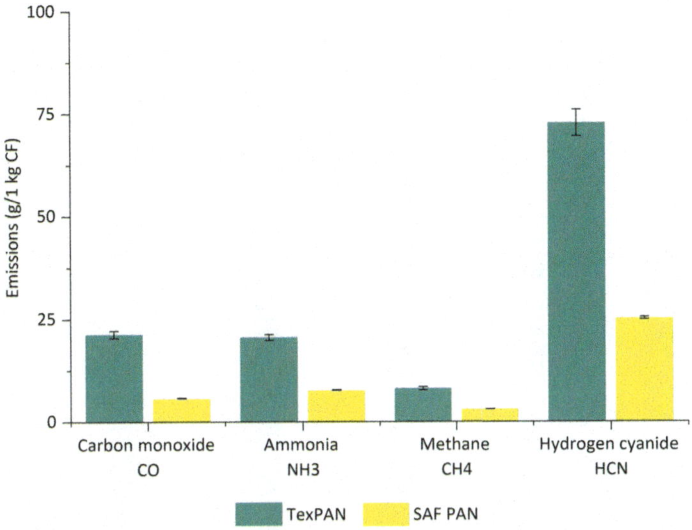

Fig. 7.9 Gaseous emissions normalized per kg carbon fibre for PAN and TexPAN. Reproduced from [1] with permission from Elsevier © Elsevier 2023

higher temperature profile for TexPAN, although earlier findings [41] suggest that small temperature changes only have a very limited effect on energy demand. Overall, the largest contributing processing step for both materials is the stabilization. As air is continuously drawn into the ovens to facilitate the oxidation reactions, the necessary power to keep temperatures stable is relatively greater than in carbonization furnaces where the atmosphere is oxygen-free and changes at a lower rate only to accommodate the removal of off-gases accrued during the respective steps [77].

For the purpose of comparison, the production of CF from LigniCel was simulated in terms of energy demand at continuous production scale. The calculation was based on a combination of processing details available from literature from benchtop experimental evaluation [36, 40, 50] and the known energy demand profile for the carbon fibre research line used in the two PAN-based trials. The results in Fig. 7.11 show a comparable energy demand for LigniCel and PAN fibre during stabilization due to a similar stabilisation time (61 vs. 58min). However, the long processing time during carbonization raises the energy demand significantly. Low temperature carbonization requires almost double the time of the already slower TexPAN process and high temperature carbonization takes almost 20 min compared to 4 and 2.4 min, for TexPAN and SAF PAN respectively.

The manufacturing scenario for LigniCel therefore requires a slightly different design than used during the PAN-based trials. For optimized LigniCel-processing, the residence time in heat in stabilization would be slightly shorter, the low temperature furnace would be a similar size, while the high temperature furnace would need to be 3 times as long to accommodate the estimated time in heat reported in Table 7.1, while running at lower manufacturing speeds than SAF PAN or TexPAN. Figure 7.11

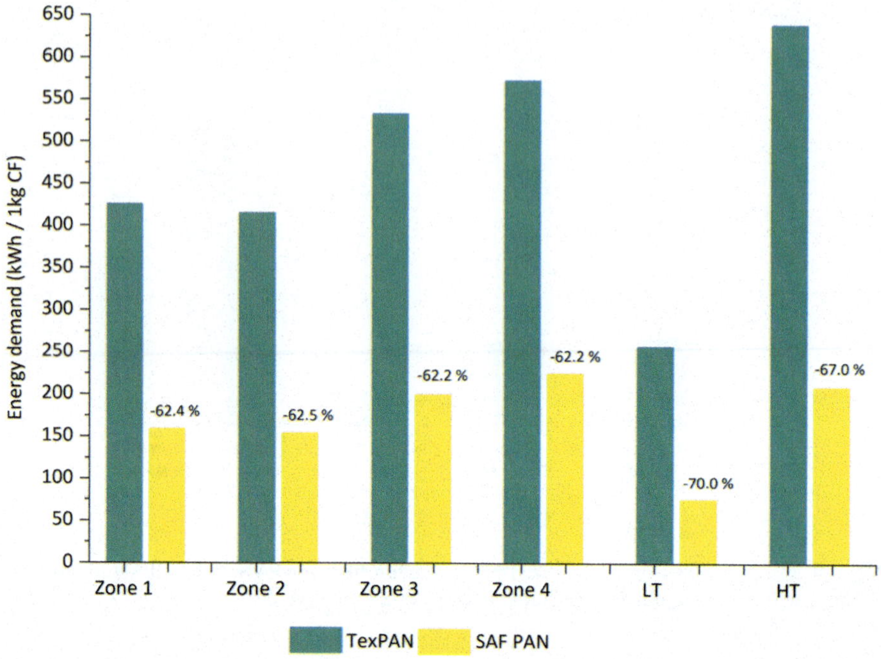

Fig. 7.10 Comparison of energy demand for SAF PAN and TexPAN. Reproduced from [1] with permission from Elsevier © Elsevier 2023

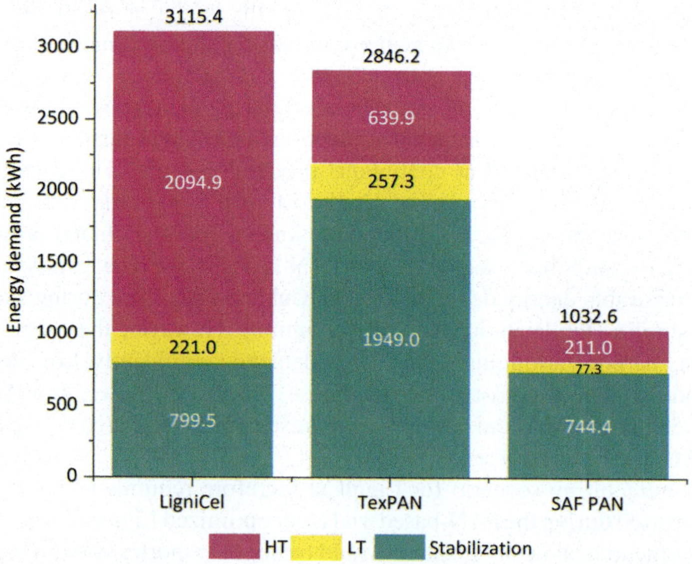

Fig. 7.11 Energy demand comparison for 1kg of carbon fibre based on three different precursors. Reproduced from [1] with permission from Elsevier © Elsevier 2023

shows that the overall energy demand is more than 300% that of PAN and the material also requires slightly more energy than TexPAN. Even though longer carbonization times and higher temperatures are required, the significantly reduced stabilization time makes up part of the higher energy usage downstream in the conversion process.

7.4.4 Cost Comparison and Impact on Commercial Viability

The data obtained for the two PAN-based precursor materials as well as the reference information gathered from literature for the LigniCel was used to compare the materials and the respective CF manufacturing processes in terms of cost. To streamline the assessment and highlight process and material influences, it was assumed that other than the material and energy related costs, all other expenses stay the same. As the two PAN-processes were analysed using the same manufacturing equipment, any non-material or directly process related costs can be assumed to be constant and not to have an influence on the comparison. Even though the LigniCel process design is slightly different, the overall necessary equipment, labour and auxiliary materials were also assumed to not significantly influence the overall cost compared to the other two processes. Based on previous research [11, 24, 33, 78] an average total price of 1kg of carbon fibre of 21.5 USD was used as a reference price for SAF PAN-based carbon fibre with the following relative cost contributions: precursor 53.2%, energy 7.1% and other contributions including labour, depreciation 39.7% [11, 24]. Consequently, over 60% of the total cost of carbon fibre are only controlled by the choice of precursor and energy intensity of the respective process (Fig. 7.12a). Based on a total cost of 21.5 USD for 1 kg of commercial PAN-based carbon fibre, the precursor cost can be deduced to be 11.44 USD and the necessary energy would contribute a further 1.53 USD. These numbers were subsequently used as a reference of cost distribution for TexPAN and LigniCel carbon fibres. Percentage disparities in precursor cost and energy for TexPAN and LigniCel were then applied relative to this cost breakdown. The results are shown in Fig. Fig. 7.12b, c.

For TexPAN, the cost ratio is significantly different to SAF PAN for two reasons. The precursor price was assumed to be significantly lower, with a cost of 5 USD per kg CF in line with other publications [11, 33, 79]. Due to the slower processing speeds however the contribution of energy-related costs raised from 7.1% to 23.8% or 4.2 USD. All other costs were assumed to be similar as the same equipment and process were used. Other publications [11, 33, 79] typically also estimate cost reductions for TexPAN-based carbon fibres, with some variability in the final price. This can be largely attributed to assumptions made such as SAF PAN reference prices, market prices for precursors and if and where improvements or disadvantages (such as higher energy demand) are attributed. The LigniCel-based CF summed up to the lowest overall cost of the three materials with 16.5 USD, almost 25% less than PAN-based carbon fibres. The contribution of the precursor was the lowest with a cost of only 3.4 USD per kg CF, while energy contributed a similar amount (28%) as for TexPAN. As Lignin is a by-product in other processes such as paper-production,

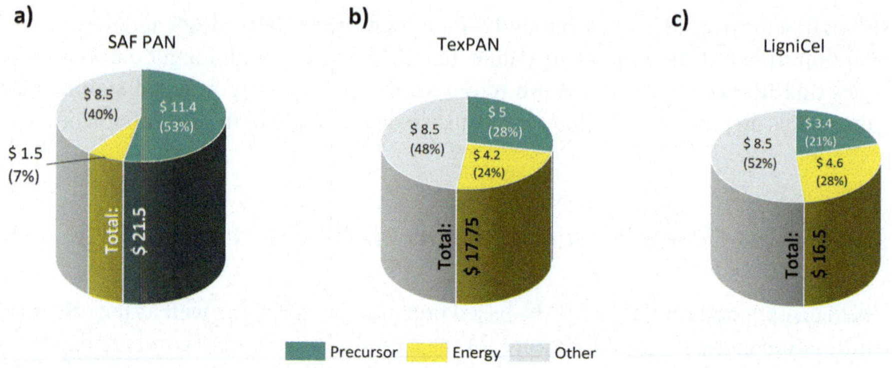

Fig. 7.12 Cost comparison of PAN-, TexPAN- and LigniCel-based carbon fibres per kg. Reproduced from [1] with permission from Elsevier © Elsevier 2023

raw material prices are very low [11, 80]. Furthermore, spinning the material does not require expensive solvents such as DMSO, however ionic liquids are often used to dissolve the cellulose, possibly impacting the cost negatively [50, 81]. However, these factors more than counterbalance the higher energy demand, particularly in the carbonization part. The assumption that all other costs are constant may introduce a low level of uncertainty, the overall trend however was clear and not thought to be affected. While calculations for the PAN-based materials used energy and material data gathered on a continuous manufacturing line, calculations for LigniCel are solely based on literature data and estimations made, based thereon. Furthermore, while LigniCel has clear advantage in terms of cost and environmental-friendliness as a bio-based material, mechanical properties are still nowhere near the values achievable with textile PAN or more so special grade PAN.

7.5 Conclusion

For the first time, this work compares three prominent precursor alternatives for carbon fibre manufacture under environmental and economic aspects, a SAF PAN, textile PAN and a more sustainable Lignin-Cellulose blend. The findings for the commercial SAF PAN in terms of material properties, energy demand, environmental impact and cost agree with values reported elsewhere. The textile PAN precursor, a lower cost alternative has never been reported to be analysed at this scale. The research conducted in this study provides indications for pathways to industrial processing of this material. While not yet widely used for carbon fibre manufacturing, the cost of textile PAN precursor is estimated to be approximately 56% lower than that of typically used special grade PAN. Novel findings in this chapter however suggest that processing costs during carbon fibre manufacture are significantly higher, as much as 176% in total, compared to PAN at the same scale and using equivalent equipment. This discrepancy is mainly due to significantly slower processing speeds and roughly 13% lower yield owing to the less process-specific fibre chemistry. The lower conversion rate also entails a higher process related environmental impact, as significantly more gasses are emitted, as high as 270% for HCN. To become a viable alternative for specialized precursor material, the processing of textile PAN needs to be improved, specifically achievable yield has to be increased as well as the tolerable processing speed. Another factor is the achievable material properties. While textile PAN-based carbon fibres can reach intermediate levels of properties, high strength or high modulus applications still require special grade precursor. The widely theoretical analysis of LigniCel showed that the material has a few significant shortcomings, which currently still inhibit commercial application. Most significantly the achievable material properties cannot match those of PAN-based fibres or even textile PAN based fibres. Furthermore, research is still in an early phase to continuously process the material on anything larger than benchtop scale. However, as a bio-based material, the environmental impact is significantly better from the start than any fossil derived material. Moreover, as a low value by-product of other processes, the low material cost is unrivalled. This study outlines basic material properties and details the cost of LigniCel-derived carbon fibres, based on a mix of real-world data and theoretical calculations, for the first time showing the potential advantages of lignin-cellulose blends as a carbon fibre precursor in greater detail. With a total cost of approximately 16 USD per kg, LigniCel-based carbon fibres are almost 25% lower in cost than PAN-based material, further increasing the competitiveness of carbon fibres compared to more traditional materials, leading to a more widespread application of the material in various industries. With intensive research ongoing to scale up LigniCel processing and improving mechanical properties, the material shows a promising pathway towards less expensive, more environmentally friendly carbon fibres.

References

1. Groetsch T, Maghe M, Creighton C, Varley RJ (2023) Economic and environmental effects of precursor variation in a continuous carbon fibre manufacturing process (in English). J Ind Eng Chem 127:554–566. https://doi.org/10.1016/j.jiec.2023.07.041
2. Morgan P (2005)Carbon fibers and their composites. CRC press, ch. 1174, pp 185–259.
3. Park S-J (2015) Carbon Fibers (Springer Series in Materials Science). Springer Science+Business Media
4. Qin Y (2016) 3—A brief description of textile fibers. In: Qin Y (ed) Medical textile materials. Woodhead Publishing, pp 23–42
5. Kaur J, Millington K, Smith S (2016) Producing high-quality precursor polymer and fibers to achieve theoretical strength in carbon fibers: a review. J Appl Polym Sci 133(38):43963. https://doi.org/10.1002/app.43963
6. Rangarajan P et al (2002) Effect of comonomers on melt processability of polyacrylonitrile. J Appl Polym Sci 85(1):69–83. https://doi.org/10.1002/app.10655
7. Tsai J-S, Lin C-H (1991) Effect of comonomer composition on the properties of polyacrylonitrile precursor and resulting carbon fiber. J Appl Polym Sci 43(4):679–685. https://doi.org/10.1002/app.1991.070430405

8. Furushima Y, Nakada M, Takahashi H, Ishikiriyama K (2014) Study of melting and crystallization behavior of polyacrylonitrile using ultrafast differential scanning calorimetry. Polymer 55(13):3075–3081. https://doi.org/10.1016/j.polymer.2014.05.015
9. Ju A, Guang S, Xu H (2013) Effect of comonomer structure on the stabilization and spinnability of polyacrylonitrile copolymers. Carbon 54:323–335. https://doi.org/10.1016/j.carbon.2012.11.044
10. Cho DW, Hong SC (2019) Synergistic effect of comonomers on the thermal oxidative stabilization of polyacrylonitrile copolymers for carbon materials. Polym Degrad Stab 161:191–197. https://doi.org/10.1016/j.polymdegradstab.2019.01.027
11. Choi D, Kil H-S, Lee S (2019) Fabrication of low-cost carbon fibers using economical precursors and advanced processing technologies. Carbon 142:610–649. https://doi.org/10.1016/j.carbon.2018.10.028
12. Gupta AK, Paliwal DK, Bajaj P (1995) Effect of an acidic comonomer on thermooxidative stabilization of polyacrylonitrile. J Appl Polym Sci 58(7):1161–1174. https://doi.org/10.1002/app.1995.070580710
13. Mathur RB, Bahl OP, Sivaram P (1992) Thermal degradation of polyacrylonitrile fibres. Curr Sci 62(10):662–669
14. Gupta AK, Paliwal DK, Bajaj P (1998) Melting behavior of acrylonitrile polymers. J Appl Polym Sci 70(13):2703–2709. https://doi.org/10.1002/(sici)1097-4628(19981226)70:13%3c2703::Aid-app15%3e3.0.Co;2-2
15. Al Aiti M, Jehnichen D, Fischer D, Brünig H, Heinrich G (2018) On the morphology and structure formation of carbon fibers from polymer precursor systems. Prog Mater Sci 98:477–551. https://doi.org/10.1016/j.pmatsci.2018.07.004
16. Maghe M et al (2016) Using ionic liquids to reduce energy consumption for carbon fibre production. J Mater Chem A 4(42):16619–16626. https://doi.org/10.1039/c6ta06842a
17. Paulauskas FL, Sherman DM (2010) Apparatus and method for stabilization or oxidation of polymeric materials. United States Patent 7649078 Patent Appl. 11/391615
18. Paulauskas FL, Naskar AK, Bonds A (2015) Advanced oxidation method for producing high-density oxidized polyacrylonitrile fibers. United States Patent US20120322332A1
19. Naskar AK et al (2017) Magneto-carbonization method for production of carbon fiber, and high performance carbon fibers made thereby. United States Patent Appl 13/833834
20. Liu J, Xiao S, Shen Z, Xu L, Zhang L, Peng J (2018) Study on the oxidative stabilization of polyacrylonitrile fibers by microwave heating. Polym Degrad Stab 150:86–91. https://doi.org/10.1016/j.polymdegradstab.2018.02.017
21. Elagib THH, Hassan EAM, Fan C, Han K, Yu M (2019) Microwave pre-oxidation for polyacrylonitrile precursor coated with nano-carbon black. Polym Eng Sci 59(3):457–464. https://doi.org/10.1002/pen.24943
22. Sunil S, Abhilas JK, Kumar A, Shukla HK (2019) Oxidative stabilization studies on pretreated polyacrylonitrile precursor fiber suitable for carbon fiber production. In: Presented at the international conference on inventive material science applications: ICIMA 2019
23. Dang W et al (2020) Structural transformation of polyacrylonitrile (PAN) fibers during rapid thermal Pretreatment in Nitrogen Atmosphere. Polymers (Basel) 12(1):63. https://doi.org/10.3390/polym12010063
24. Nunna S, Blanchard P, Buckmaster D, Davis S, Naebe M (2019) Development of a cost model for the production of carbon fibres. Heliyon 5(10):e02698. https://doi.org/10.1016/j.heliyon.2019.e02698
25. Warren C, Naskar A (2012) Lower cost carbon fiber precursors. In: Presentation at 2012 DOE hydrogen and fuel cells program and vehicle technologies program annual merit review and peer evaluation meeting, Arlington, VA, USA, 2012, vol 16
26. Yoo SH et al (2017) Facile method to fabricate carbon fibers from textile-grade polyacrylonitrile fibers based on electron-beam irradiation and its effect on the subsequent thermal stabilization process. Carbon 118:106–113. https://doi.org/10.1016/j.carbon.2017.03.039
27. Ennis BL et al (2019) Optimized carbon fiber composites in wind turbine blade design. Sandia National Lab.(SNL-NM), Albuquerque, NM (United States)

28. Eslami Farsani R, Shokuhfar A, Sedghi A (2006) Stabilization of commercial polyacrylonitrile fibres for fabrication of low-cost medium-strength carbon fibres. e-Polymers 6(1). https://doi.org/10.1515/epoly.2006.6.1.1
29. Khan H, Kaur J, Naebe M, Hutchinson S, Varley RJ (2022) Continuous, pilot-scale production of carbon fiber from a textile grade PAN polymer. Mater Today Commun 31. https://doi.org/10.1016/j.mtcomm.2022.103231.
30. Hiremath N, Young S, Ghossein H, Penumadu D, Vaidya U, Theodore M (2020) Low cost textile-grade carbon-fiber epoxy composites for automotive and wind energy applications. Compos Part B: Eng 198. https://doi.org/10.1016/j.compositesb.2020.108156
31. König S et al (2021) High-performance carbon fibers prepared by continuous stabilization and carbonization of electron beam-irradiated textile grade polyacrylonitrile fibers. Macromolecular Mater Eng 306(12). https://doi.org/10.1002/mame.202100484.
32. Wheatley A, Warren D, Das S (2013) Low-cost carbon fibre: applications, performance and cost models. In: Advanced composite materials for automotive applications, pp 405–434
33. Paulauskas F et al (2009) Novel Precursor materials and approaches for producing lower cost carbon fiber for high volume industries. Automot Appl and Recycl
34. Warren C (2011)Low cost carbon fiber overview. Oak Ridge National Laboratory, Oak Ridge, Tennessee
35. Bonds T (2020) Low cost, high volume, carbon fiber precursor for plasma oxidation
36. Bengtsson A et al (2018) Improved yield of carbon fibres from cellulose and kraft lignin. Holzforschung 72(12):1007–1016. https://doi.org/10.1515/hf-2018-0028
37. Trogen M et al (2021) Cellulose-lignin composite fibres as precursors for carbon fibres. Part 1—Manufacturing and properties of precursor fibres. Carbohydr Polym 252:117133. https://doi.org/10.1016/j.carbpol.2020.117133
38. Bajpai P (2017)Lignin as a raw material for carbon fibre. In: Carbon Fibre from Lignin. SpringerBriefs in Materials, Springer, Chapter 5, pp 29–32
39. Patil ND, Tanguy NR, Yan N (2016) Lignin interunit linkages and model compounds. In: Lignin in polymer composites, pp 27–47
40. Bengtsson A et al (2022) Continuous stabilization and carbonization of a lignin-cellulose precursor to carbon fiber. ACS Omega 7(19):16793–16802. https://doi.org/10.1021/acsomega.2c01806
41. Groetsch T, Maghe M, Creighton C, Varley RJ (2023) Environmental, property and cost impact analysis of carbon fibre at increasing rates of production. J Cleaner Produc 382. https://doi.org/10.1016/j.jclepro.2022.135292
42. Le ND, Trogen M, Ma Y, Varley RJ, Hummel M, Byrne N (2020) Cellulose-lignin composite fibers as precursors for carbon fibers: Part 2 - The impact of precursor properties on carbon fibers. Carbohydr Polym 250:116918. https://doi.org/10.1016/j.carbpol.2020.116918
43. Baker DA, Gallego NC, Baker FS (2012) On the characterization and spinning of an organic-purified lignin toward the manufacture of low-cost carbon fiber. J Appl Polym Sci 124(1):227–234. https://doi.org/10.1002/app.33596
44. Zhang M, Ogale AA (2014) Carbon fibers from dry-spinning of acetylated softwood kraft lignin. Carbon 69:626–629. https://doi.org/10.1016/j.carbon.2013.12.015
45. Baker DA, Rials TG (2013) Recent advances in low-cost carbon fiber manufacture from lignin. J Appl Polym Sci 130(2):713–728. https://doi.org/10.1002/app.39273
46. Li Z, Rosli RAB, Xiang Y, Ye E, Loh XJ (2018) Carbon precursor from lignin: methods and applications. In: Functional materials from lignin. Sustainable Chemistry Series, pp 121–152
47. Dai Z, Shi X, Liu H, Li H, Han Y, Zhou J (2018) High-strength lignin-based carbon fibers via a low-energy method. RSC Adv 8(3):1218–1224. https://doi.org/10.1039/c7ra10821d
48. Le N-D, Trogen M, Ma Y, Varley RJ, Hummel M, Byrne N (2020) Understanding the influence of key parameters on the stabilisation of cellulose-lignin composite fibres. Cellulose 28(2):911–919. https://doi.org/10.1007/s10570-020-03583-y
49. Le N-D, Trogen M, Varley RJ, Hummel M, Byrne N (2020) Effect of boric acid on the stabilisation of cellulose-lignin filaments as precursors for carbon fibres. Cellulose 28(2):729–739. https://doi.org/10.1007/s10570-020-03584-x

50. Bengtsson A, Bengtsson J, Sedin M, Sjöholm E (2019) Carbon fibers from lignin-cellulose precursors: effect of stabilization conditions. ACS Sustainable Chem Eng 7(9):8440–8448. https://doi.org/10.1021/acssuschemeng.9b00108
51. Nunna S et al (2019) Time dependent structure and property evolution in fibres during continuous carbon fibre manufacturing. Materials (Basel), 12(7):1069. https://doi.org/10.3390/ma12071069
52. Groetsch T et al (2021)A modular LCA/LCC-modelling concept for evaluating material and process innovations in carbon fibre manufacturing. Procedia CIRP 98:529–534. https://doi.org/10.1016/j.procir.2021.01.146
53. Jin X, Feng C, Creighton C, Hameed N, Parameswaranpillai J, Salim NV (2021) On the structural evolution of textile grade polyacrylonitrile fibers during stabilization and carbonization: towards the manufacture of low-cost carbon fiber. Polymer Degradation and Stability 186. https://doi.org/10.1016/j.polymdegradstab.2021.109536
54. Park S-J, Heo G-Y (2015) Precursors and Manufacturing of Carbon Fibers. In: Park S-J (ed) Carbon Fibers. Springer Series in Materials Science. Springer Netherlands, Dordrecht , Chapter 2, pp 31–66
55. Khayyam H et al (2020) PAN precursor fabrication, applications and thermal stabilization process in carbon fiber production: experimental and mathematical modelling. Prog Mater Sci 107:100575. https://doi.org/10.1016/j.pmatsci.2019.100575
56. Donnet J-B, Chand Bansal R (1998) Carbon fibers. CRC Press ch. 572, pp 1–370
57. Bengtsson A, Hecht P, Sommertune J, Ek M, Sedin M, Sjöholm E (2020) Carbon fibers from lignin-cellulose precursors: effect of carbonization conditions. ACS Sustainable Chem Eng 8(17):6826–6833. https://doi.org/10.1021/acssuschemeng.0c01734
58. Groetsch T et al (2021) Gas emission study of the polyacrylonitrile-based continuous pilot-scale carbon fiber manufacturing process. Ind Eng Chem Res. https://doi.org/10.1021/acs.iecr.1c02253
59. Takaku A, Hashimoto T, Miyoshi T (1985) Tensile properties of carbon fibers from acrylic fibers stabilized under isothermal conditions. J Appl Polym Sci 30(4):1565–1571. https://doi.org/10.1002/app.1985.070300421
60. Bajaj P, Roopanwal AK (1997) Thermal stabilization of acrylic precursors for the production of carbon fibers: an overview. J Macromolecular Sci, Part C: Polymer Rev 37(1):97–147. https://doi.org/10.1080/15321799708014734
61. Ge Y, Fu Z, Zhang M, Zhang H (2020) The role of structural evolution of polyacrylonitrile fibers during thermal oxidative stabilization on mechanical properties. J Appl Polym Sci 138(1):49603. https://doi.org/10.1002/app.49603
62. Jin Huang SF, Gan L (2019) Structure and characteristics of lignin. In: Jin Huang SF, Gan L (eds) Lignin chemistry and applications, pp 25–50
63. Byrne N, De Silva R, Ma Y, Sixta H, Hummel M (2018) Enhanced stabilization of cellulose-lignin hybrid filaments for carbon fiber production. Cellulose (Lond) 25(1):723–733. https://doi.org/10.1007/s10570-017-1579-0
64. Wypych A (2018) Lignin. In: Databook of adhesion promoters
65. Sun CC (2005) True density of microcrystalline cellulose. J Pharmaceut Sci 94(10):2132–2134. https://doi.org/10.1002/jps.20459
66. Fitzer E, Frohs W, Heine M (1986) Optimization of stabilization and carbonization treatment of PAN fibres and structural characterization of the resulting carbon fibres. Carbon 24(4):387–395. https://doi.org/10.1016/0008-6223(86)90257-5
67. Dumanlı AG, Windle AH (2012) Carbon fibres from cellulosic precursors: a review. J Mater Sci 47(10):4236–4250. https://doi.org/10.1007/s10853-011-6081-8
68. Xue TJ, McKinney MA, Wilkie CA (1997) The thermal degradation of polyacrylonitrile. Polym Degrad Stab 58(1–2):193–202. https://doi.org/10.1016/s0141-3910(97)00048-7
69. Kennedy JP, Fontana CM (1959) The thermal degradation of polyacrylonitrile. J Polym Sci 39(135):501–506. https://doi.org/10.1002/pol.1959.1203913542
70. Hanna SB, Yehia AA, Ismail MN, Khalaf AI (2012) Preparation and characterization of carbon fibers from polyacrylonitrile precursors. J Appl Polym Sci 123(4):2074–2083. https://doi.org/10.1002/app.34704

71. Ji M, Wang C, Bai Y, Yu M, Wang Y (2007) Structural evolution of polyacrylonitrile precursor fibers during preoxidation and carbonization. Polym Bull 59(4):527–536
72. Wang PH, Yue ZR, Li RY, Liu J (1995) Aspects on interaction between multistage stabilization of polyacrylonitrile precursor and mechanical properties of carbon fibers. J Appl Polym Sci 56(2):289–300. https://doi.org/10.1002/app.1995.070560221
73. Chung DDL (1994) Carbon Fiber Composites (no. 1). Heinemann, Newton, MA, p 209
74. Frank E, Hermanutz F, Buchmeiser MR (2012) Carbon fibers: precursors, manufacturing, and properties. Macromol Mater Eng 297(6):493–501. https://doi.org/10.1002/mame.201100406
75. Mansouri N-EE, Salvadó J (2006) Structural characterization of technical lignins for the production of adhesives: application to lignosulfonate, kraft, soda-anthraquinone, organosolv and ethanol process lignins. Ind Crops Prod 24(1):8–16. https://doi.org/10.1016/j.indcrop.2005.10.002
76. Sedghi A, Farsani RE, Shokuhfar A (2008) The effect of commercial polyacrylonitrile fibers characterizations on the produced carbon fibers properties. J Mater Process Technol 198(1–3):60–67. https://doi.org/10.1016/j.jmatprotec.2007.06.052
77. Dér A et al (2021) Modelling and analysis of the energy intensity in polyacrylonitrile (PAN) precursor and carbon fibre manufacturing. J Clean Prod 303:127105. https://doi.org/10.1016/j.jclepro.2021.127105
78. Ellringmann T, Wilms C, Warnecke M, Seide G, Gries T (2015) Carbon fiber production costing: a modular approach. Text Res J 86(2):178–190. https://doi.org/10.1177/0040517514532161
79. Warren CD (2013) Lightweighting composites and lower cost carbon fiber. Oak Ridge National Laboratory, US Department of Energy (March 2013)
80. Baker F, Gallego N, Baker D (2010) Low cost carbon fiber from renewable resources. EERE, US Dept of Energy Project ID# lm_03_baker
81. Olsson C, Sjöholm E, Reimann A (2017) Carbon fibres from precursors produced by dry-jet wet-spinning of kraft lignin blended with kraft pulps. Holzforschung 71(4):275–283. https://doi.org/10.1515/hf-2016-0189

Chapter 8
Conclusions and Potential Areas for Future Work

Abstract This chapter summarizes the work presented in this book, starting with the development of a streamlined life cycle assessment and life cycle costing framework tailored for carbon fibre manufacturing, addressing both high costs and environmental impacts. Subsequently the effect of how manufacturing conditions impact material properties, energy use, emissions, and overall costs is discussed. Potential savings through optimized processing scenarios are identified but trade-offs in mechanical performance at higher speeds are outlined. Additionally, the evaluated alternative precursor materials are discussed, with lower-cost options like textile PAN and bio-based blends shown to offer some economic benefits, but they also present challenges in energy consumption and material quality.

8.1 Summary

Carbon fibre manufacture is an expensive and energy intensive process that consequently has a high environmental impact. While the material and related production processes have been researched and improved over many years, the secretive nature of the carbon fibre industry has limited the publicly available knowledge. Any progress in development is therefore, among other reasons, incremental. Available process and material analysis studies often lack critical detail to compare their outcome and the scope varies widely. A streamlined methodology to assess carbon fibre manufacture may provide valuable insights into improvement potential for process and material optimization pathways. This can help overcome hurdles in material and process development and uncover hot spots where improvements or changes yield the greatest impacts (Fig. 8.1).

In this study a streamlined assessment concept was developed based on life cycle assessment and life cycle costing principles. Both methods are well-established tools to analyse products and processes with a holistic perspective. The combination of an environmental and cost perspective can address two key concerns associated with carbon fibre material, the high price and significant environmental burden caused

T. Groetsch, *Enhancing Carbon Fibre Production Through Life Cycle Analysis*, Springer Theses, https://doi.org/10.1007/978-3-031-70578-6_8

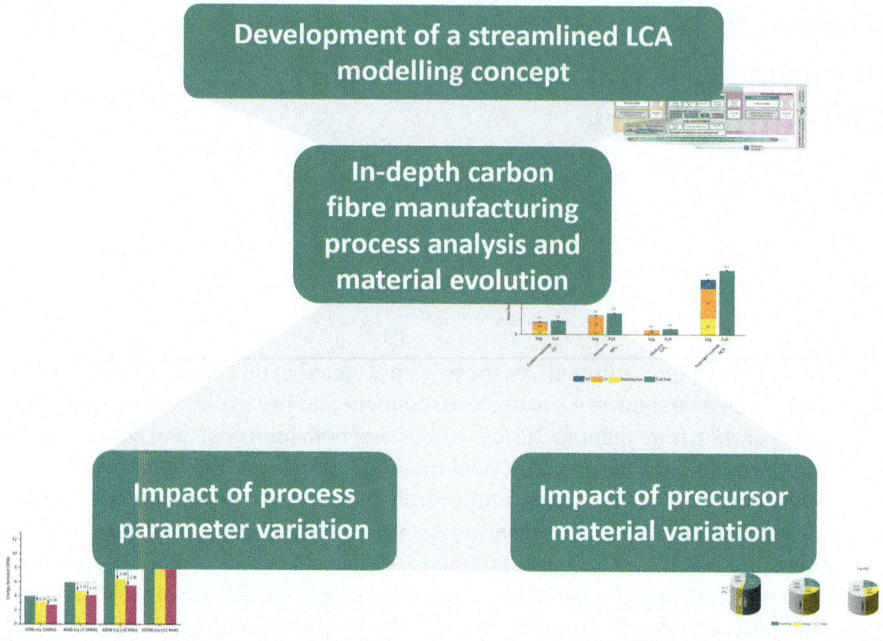

Fig. 8.1 Approach to uncover improvement potentials in carbon fibre manufacturing by application of a streamlined LCA/LCC modelling concept

by the energy and raw-material intensive production. The methodologies 'Environmental Impact Analysis' and 'Cost Analysis' were combined and the overall approach was tailored to the carbon fibre manufacturing process. Any steps and parts of the standardized methodology that were assessed to have no significant contribution, were excluded to keep the concept focused on key elements and straightforward to use. Manufacturing process steps and equipment was grouped into few sections for a clear view and to further increase usability. To showcase the concept applicability an initial exemplary assessment was performed with well-established carbon fibre manufacturing process parameters and known source material.

With publicly available process assessments often lacking detail or omitting parts an in-depth holistic analysis of the manufacturing process was performed, encompassing not only the process, related energy demand, cost and emissions, but also examine how the transformation process manifests in the material. Using a commercial precursor fibre, carbon fibres were produced on the Carbon Nexus research line, a scaled down version of industrial processing. During the campaign all relevant process data was recorded, emissions were measured and fibre samples were taken after each manufacturing step to be analysed for their structural and mechanical properties. The work contributed to an improved understanding of the relationship between processing conditions, material structure and properties, energy demand profile, emissions composition, and by extension environmental impact and cost.

With this thorough assessment as a baseline, significant process parameters were analysed in more detail for the purpose of identification of energy and cost saving potential. Four different processing scenarios were designed that fulfill the requirement of producing carbon fibre of equal quality and properties. Four manufacturing experiments with varying carbon fibre throughput were carefully optimized to produce the desired material, while running the equipment at the best possible conditions in terms of energy demand, material utilisation and emission generation. While increasing processing speeds generally improved the overall cost and reduced environmental impact, a trade-off with reduction in mechanical performance became evident for very high speeds. The findings were subsequently used to predict the energy and cost saving potentials at larger, commercial production scales.

A further study (Chap. 7) was dedicated to developing a better understanding of the effect of choice of precursor material on the material-specific process parameters during the carbonisation process and associated environmental impact and cost. Three different materials were processed and analysed. The baseline precursor, once again, was a commercial PAN fibre with a well-established set of processing parameters. The second material was a textile PAN fibre, which is generally produced at lower cost at commercial quantities but lacks processability enhancing comonomers. The third material was an experimental bio-based lignin-cellulose blend. With increasing awareness of resource depletion in combination with the high environmental burden of fossil-based precursors, such materials promise a benefits in terms of cost and environmental impact. As this material was still in a highly experimental phase of development, processing on the same scale as the other two materials was not possible. Any missing information was therefore substituted with estimations based on small scale experiments and literature. While both the textile and bio-based carbon fibres provide the benefit of lower precursor-associated costs, their processing is slower and in case of the bio-based material significantly more challenging in terms of achieving relevant mechanical properties and a high material yield. The outcome of this work showed that positive effect associated with the low cost of the two alternative precursors are to some extent reduced by processing related negative impacts on energy consumption and environment.

8.2 Conclusions

The findings of this work contribute to a better understanding of the carbon fibre manufacturing process from a life cycle perspective. The work further demonstrates the relationship of key aspects of the production process and material properties based on streamlined studies of various manufacturing conditions and material types.

8.2.1 Development of a Streamlined Carbon Fibre Manufacturing Assessment Concept

Based on well-established and standardized LCA and LCC methodologies, a streamlined assessment concept for carbon fibre manufacture was developed and applied to a set of processing conditions as a proof of concept in Chap. 4. Key findings in developing and applying this methodology were:

- Due to the high complexity of the underlying methodology and the necessary expert knowledge to perform full scale LCA or LCC studies, streamlining the concept and only specific aspects increases application potential while decreasing risk of procedural errors.
- A clear, modular structure can help increase transparency, ease of use and comparability with other processes or assessments. Based on published assessment results in the field of carbon fibre manufacture, the comparison with other LCA studies related to carbon fibre manufacture is challenging as the scope of analyses vary and important information is often not outlined in detail.
- The modularity of the concept allows to include or exclude stages of the manufacturing process chain, such as precursor production or composite fabrication. Applying the same streamlined approach with predefined, aggregated steps ensures transparency and consistency while keeping the required knowledge and data to a minimum.
- Even though the concept includes the aggregation of results into impact categories and summarized costs, this step was found to be not required to compare the impact of process-related changes, as energy related emissions are the biggest contributor to each category and linearly correlated to the categories.

Overall the merit of a streamlined LCA/LCC approach for early phases of process and material development has been demonstrated in this work. The concept facilitated the evaluation of various processing conditions and different precursor material with manageable effort and helped uncover improvement potential related to the process as well as the precursor.

8.2.2 Analysis of Carbon Fibre Manufacturing Process and Material Evolution

Based on the above-mentioned assessment concept, the carbon fibre manufacturing process was studied in depth on a granular step-by-step basis. Correlations between process parameters, emission composition, material properties and structure were characterized and outlined on a continuous manufacturing scale in detail in Chap. 5. Findings of this work can be summarized as follows:

- The main emissions (not normally present in the atmosphere) generated during carbonization of PAN-based precursor were HCN, NH_3, CO and CH_4. HCN was by far the biggest contributor to the emission composition and was mostly generated during low temperature carbonization.
- In general most gases were generated in the low temperature carbonization section, corresponding to a high observed mass loss of the fibre. During stabilization the exhaust volume was observed to be the highest, but with only low concentrations of the analysed gases present. As most non-carbon elements have been removed during stabilization and LT treatment, low levels of emissions were observed in the high temperature carbonization.
- Elemental analysis of the fibres taken along the manufacturing process revealed that during stabilization the amount of N, C and H slightly decreased while the uptake of O into the chemical structure significantly increased. During carbonization, especially in the HT stage, a significant reduction in non-carbonaceous elements was observed. This trend very much correlated with the gaseous emissions generated during the respective manufacturing steps.

For the first time, this work generated a holistic assessment approach to the carbon fibre manufacturing process including the identification of assessment criteria arising from process parameter and emission contributions on a detailed production step level, but also under consideration of processing parameter-property-relationships.

8.2.3 Impact of Process Parameter Variation

Applying the developed streamlined assessment concept and findings from the gas emission study in Chap. 5, the impact of processing throughput on energy demand and evolved emissions at each carbon fibre production step was analysed in Chap. 6. A commercial grade PAN precursor material was processed at four different manufacturing speeds, 10, 20, 30 and 40 m/h. Each production scenario was extensively optimized to yield the same quality and grade of carbon fibre material. These results were used to predict the impact of these changes on commercial production quantities and potential energy and cost savings. The main findings included:

- Even though the temperatures were slightly increased with faster processing speed, small temperature changes in the range of 5–10 °C did have a visible impact on the overall high level of energy usage
- The range in which the manufacturing process can be optimized towards higher processing speeds is limited. The study clearly showed that, while the desired material properties can be achieved at very high speeds, the consistency of the material decreases.
- The aggressive treatment at very high speeds led to lower carbon yield due to the short residence time the fibre remained in the carbonisation furnaces. The lower carbon yield had a major effect on the energy demand calculated per kg carbon fibre (MJ/kg) and in turn on the environmental impact and cost.

- Scaled-up to commercial production quantities, faster processing showed to have significant benefits, both in terms of energy and cost savings. The reduced yield at very high speeds, however, eliminates any benefit in energy savings and entails higher overall costs.

This work clearly highlighted the benefits of a well optimized production process. While accelerating carbon fibre throughput can achieve significant energy and cost savings on a commercial scale, the maximum production speed is limited by material chemistry and resultant material properties and their consistency. In this chapter, the overall carbon yield was found to be the most important factor for energy and cost-effective production of carbon fibre, followed by the material throughput and the temperature profile.

8.2.4 Impact of Precursor Material Variation

The choice of source materials can have a significant impact on processing conditions, energy demand, cost and emissions in a production environment. In light of this, Chap. 7 assesses the cost benefit and environmental impact of three different precursor materials for carbon fibre production: 1. traditionally used SAF PAN, 2. textile, low-cost PAN and 3. bio-based Lignin-Cellulose precursor. Data was derived from in-house continuous production trials for the two PAN-based materials, and small-scale experiments and literature values for the bio-based precursor, which were up-scaled for processes comparison. The assessment showed the following results:

- Findings for the reference material (1) traditional SAF PAN precursor) in terms of energy demand, emissions and cost confirm values reported elsewhere, making the material a reliable baseline for this analysis
- The textile material (2.) could be processed to achieve very similar material as the special grade PAN. Two key differences were the significantly lower processing speed and a lower overall yield
- The cost benefit of the lower priced textile PAN precursor (2.) was to some extent diminished by the slower processing speed and therefore resulted in a higher energy demand per kg carbon fibre. The overall assessment however suggested that using textile-based PAN (2.) fibre as a carbon fibre precursor has a cost advantage of about 4 USD over SAF PAN (1.). However, the analysis revealed that, due to the slower processing, the emissions and environmental impact are higher.
- With lignin being a by-product of the pulp and paper-making industry, the lignin-cellulose blend (3.) provides by far the biggest cost advantage of all analysed precursors in this study. Nevertheless, this advantage was offset by slower, more complex processing and a significantly lower carbon fibre yield of around 33%. Due to the very low cost of the precursor, however, the overall price of lignin-cellulose carbon fibres is expected to be lower than SAF PAN- and textile PAN -based carbon fibres. The assessment showed that the environmental impact of

lignin-cellulose carbon fibres is also expected to be significantly lower than the fossil-based alternatives. However, this bio-based precursor is still under development and today its material properties have not yet reached those of PAN-based fibres

This work compared three precursor alternatives for carbon fibre manufacture based on large-scale process data and literature values. The analysis showed that the use of both the textile and bio-based material may open pathways to low-cost carbon fibres. With an advancement in the development of bio-based carbon fibres towards higher mechanical performance in the future; lignin/cellulose precursors will become a significant alternative to PAN precursors due to their low price and expected lower environmental impact.

8.3 Summary

The developed gate-to-gate assessment methodology proved a useful tool for evaluating a variety of processing conditions and precursor materials for a predefined carbon fibre manufacturing process. While based on LCA and LCC principles, the streamlined nature of the concept allowed for a rapid analysis of a large number of manufacturing scenarios and identification of optimal processing conditions. At the same time the modularity enabled case-by-case adjustments to in- or exclude elements of the methodology and pivot towards specific key results, i.e. focus more on environmental impacts for one study and cost in the other. Most significantly, the aggregation in impact categories was only used initially as for the remainder of the work, results could be shown equally understandable without the conversion.

Applying the developed assessment method to study the carbon fibre manufacturing process highlighted interrelationships between the material properties, process parameters, emissions and energy demand. This in turn demonstrated the relative contribution of each step in the processing chain to the environmental impact of the material and emphasized that energy production related emissions dwarf any in-process off-gases.

The analysis of different process conditions, namely manufacturing rates, revealed that optimization towards an ideal balance of speed and aggressiveness of treatment can lead to significant energy and cost savings. The optimizations employed did not require any additional equipment or treatment and therefore showcase the importance of properly investigating the process and its limits when scaling to commercial production.

The comparison of different precursors showed that the material selection plays a key role in determining the overall environmental impact and cost of the final material. While manufacturing carbon fibre from alternative precursor materials to PAN, such as textile PAN or bio-based materials is significantly slower and more complex, benefits from the lower precursor material cost make up for additional costs incurred here. The environmental impact of textile PAN however was shown to be higher than for special grade PAN, while bio-based precursor was shown to have the biggest environmental advantage.

Overall this work contributed to a better understanding of the carbon fibre manufacturing process on a continuous, industrial scale and highlighted areas of future focus in material or process development or when upscaling benchtop scale research. The study also outlined pathways for improving the environmental impact and to lower the cost of carbon fibre.

8.4 Recommendations for Future Work

Most of the suggested future work is based on the gate-to-gate environmental and economic assessment methodology outlined in Chap. 4. The assessment method can be used to expand the study of scenarios and build an assessment results inventory database for comparison and to further improve the understanding of the process. The methodology itself can also further be enhanced by extending and defining it for additional production steps.

Topics that have merit to be investigated include:

- *The effect of tow size:* Another route of a more economic approach of manufacturing carbon fibre is the application of precursors with larger tow size. While the processing of precursor fibres with larger tow sizes is expected to have a beneficial effect on the final product cost, new processing challenges are introduced. A better understanding of the benefits and disadvantages as well as potential sweet spots may help steer research towards quicker improvements
- *The effect of annual throughput:* Similarly to tow size, the overall size of the production line may also merit further analysis. Larger equipment capable of processing more material is also expected to improve the cost structure and environmental impact on a per kg carbon fibre basis.
- *The effects of scaling the manufacture towards commercial sized production:* At Carbon Nexus there are two carbon fibre production lines of different sizes available. This unique setup enables the comparison of results at different production scales and how the line size influences production parameters, energy consumption and environmental effects. Through comparing assessment results with the same material input/output on different size production lines, relations between relevant parameters could be investigated. This should help understand scaling factors for critical production parameters better and in turn enable virtually transferring assessment results to other production lines.
- *Gas emission treatment*: Carbon fibre manufacture produces a variety of toxic gases, which have to be treated before being released into the atmosphere. The most common process to treat any generated emissions is through thermal oxidation. There are several technical concepts achieving this (i.e. direct fired thermal oxidation, regenerative thermal oxidation) and industry scale production facilities generally use different types for oxidation ovens and LT/HT-furnaces for increased

efficiency. As abatement systems require a substantial amount of energy to treat the hazardous gases, it may be beneficial to have a closer look at the abatement process and how it can be optimized.

- *Extended LCA/LCC - precursor:* One key aspect to further investigate is to expand the scope to include up- and/or downstream processes. As the production of precursor material is a very similar process to carbon fibre production in terms of continuous production process, and the precursor material price is a considerable part of the overall carbon fibre cost, it may be beneficial to include its production in extended or separate assessments. The tailoring of the assessment methodology would follow the same structure as for the carbon fibre model and be based on the already existing structure.
- *Extended LCA/LCC—recycling:* With increasing volume of carbon fibre used, recycling becomes more important. Today's recycling concepts usually employ a second use concept rather than a fully circular approach. Furthermore the approaches usually focus on the recovery of either the fibres or the matrix material through mechanical, chemical or thermal processes. For the waste material that is incurred during the production of the fibre, specifically during ramp-up/-down, no recycling concept could be found during the review. This may be due to the fact that in relation to the output volume the amount of generated waste is relatively little, nevertheless reducing waste by any amount improves the financial and environmental impact of the product. Published research on recycling usually focuses of the technical side, only limitedly considering cost or environmental impact. Comparing recycling strategies, their impact and virgin carbon fibre production may help outline requirements for recycling to be a beneficial step.

About the Author

Dr. Thomas Groetsch is a Postdoctoral Research Fellow at the Institute for Frontier Materials at Deakin University in Waurn Ponds, Australia, where he is currently working in Prof. Russell J. Varley's Composites group. Prior to his current role at Deakin University he worked as a research and validation engineer in the automotive industry in Munich, Germany (2016–2019). He gained a PhD Degree from the Deakin University, Austrialia (2023), a Master of Engineering in Automotive Engineering from the Technical University Ingolstadt, Germany (2016) and Bachelor of Engineering in Mechanical Engineering (2013) from the Technical University Deggendorf, Germany.

His current research interests cover sustainable manufacture and processing optimization of carbon fibres and composites, and bio-based composite materials. Recently, he has been focusing on the topic of high value recycling of composite materials.

T. Groetsch, *Enhancing Carbon Fibre Production Through Life Cycle Analysis*, Springer Theses, https://doi.org/10.1007/978-3-031-70578-6

The manufacturer's authorised representative in the EU is Springer Nature Customer Service Centre GmbH, Europaplatz 3, 69115 Heidelberg, Germany. If you have any concerns regarding our products, please contact ProductSafety@springernature.com

Printed and bound by CPI Group (UK) Ltd, Croydon, CR0 4YY
30/09/2026
02225604-0009